Water Resources

Water Resources: Health, Environment and Development

Edited by
Brian H. Kay

E & FN SPON
An imprint of Routledge

London and New York

First published 1999
by E & FN Spon, an imprint of Routledge
11 New Fetter Lane, London EC4P 4EE

Simultaneously published in the USA and Canada
by Routledge
29 West 35th Street, New York, NY 10001

Typeset in 10/12pt Times by Keystroke, Jacaranda Lodge, Wolverhampton
Printed and bound in Great Britain by Biddles Ltd, Guildford and King's Lynn

British Library Cataloguing in Publication Data
A catalogue record for this book is available from the British Library

Library of Congress Cataloguing in Publication Data
A catalog record for this book has been requested

ISBN 0 419 22290 1

Contents

Contributors

Dr Keith W. Bentley
Director, Environmental Health Policy
Commonwealth Department of Health and Family Services
Canberra ACT 2601
Australia

Dr Martin H. Birley
The Liverpool Health Impact Programme
Liverpool School of Tropical Medicine
Pembroke Place
Liverpool L3 5QA
UK

Mr Robert Bos
Executive Secretary, PEEM
World Health Organization
Division of Environmental Health
CH-1211 Geneva 27
Switzerland

Dr Annette K. Broom
Department of Microbiology
University of Western Australia
QEII Medical Centre
Nedlands WA 6907
Australia

Professor G. Dennis Calvert
Professor of Medicine and Public Health
Head, Medical Research Unit
University of Wollongong
PO Box 1144
Wollongong NSW 2522
Australia

Dr Joseph C. Cooney
Tennessee Valley Authority
P.O. Box 1010
Muscle Shoals, AL 35662–1010
USA
Current Address:
117 Highland Place
Sheffield, AL 35660
USA

Professor Charles E. Engel
Medical Education Unit
University College/London Medical School
46 Cleveland Street
London W1P 6DB
UK

Professor Christine E. Ewan
Pro Vice-Chancellor (Academic)
University of Wollongong
PO Box 1144
Wollongong NSW 2522
Australia

Dr Peter Furu
Senior Lecturer
Environmental Health Biologist

Danish Bilharziasis Laboratory
Jaegersborg Allé 1D
DK-2920 Charlottenlund
Denmark

Dr Mark Hearnden
Department of Primary Industries and Fisheries
GPO Box 990
Darwin NT 0801
Australia

Dr Marina Hurley
School of Australian Environmental Sciences
Griffith University
Nathan Qld 4111
Australia

Dr Arumugam Kandiah
Senior Officer
Water Resources Development and Management Service
Food and Agriculture Organization of the United Nations
Viale delle Terme di Caracalla
00100 Rome
Italy

Assoc. Professor Brian Kay
Head, Mosquito Control Laboratory
Queensland Institute of Medical Research
PO Royal Brisbane Hospital
Brisbane Qld 4029
Australia

Professor John S. Mackenzie
Head, Department of Microbiology
University of Queensland
Brisbane Qld 4072
Australia

Dr Kamolnetr Okanurak
Social and Economic Research Unit
Department of Tropical Medicine
Mahidol University
Bangkok
Thailand

Professor Richard Pearson
Director, Australian Centre for Tropical Freshwater Research
James Cook University
Townsville Qld 4811
Australia

Dr Walter Rast
Manager, Freshwater Programme
United Nations Environment Programme
PO Box 30552
Nairobi
Kenya

Assoc. Professor Richard C. Russell
Director, Dept Medical Entomology
University of Sydney
ICPMR, Westmead Hospital
Westmead NSW 2145
Australia

Mr W. Bart Snellen
International Institute for Land Reclamation and Improvement
PO Box 45
6700 AA Wageningen
The Netherlands

Dr Edward L. Snoddy (deceased)
Medical Entomologist, late of:
Environmental Research Center
Tennessee Valley Authority
PO Box 1010
Muscle Shoals AL 35662-1010
USA

Professor Santasiri Sornmani
formerly, Social and Economic Research Unit
Department of Tropical Medicine
Mahidol University
Bangkok
Thailand

Mr Kenneth J. Tennessen
Tennessee Valley Authority
PO Box 1010
Muscle Shoals AL 35662-1010
USA

Dr Jeffrey A. Thornton
International Environmental
Management Services Ltd
321 Barney Street
Waukesha WI 53186
USA

Foreword

Freedom from hunger and malnutrition, good health and a safe and stable environment are basic to sustainable development. The Heads of State and Government gathered at the World Food Summit in Rome in November 1996 considered it intolerable that more than 800 million people throughout the world, and particularly in developing countries, do not have enough food to meet their basic and nutritional needs. The participants of the Summit pledged their political will and their common and national commitment to achieving food security for all and to an ongoing effort to eradicate hunger in all countries. The growing population of the hungry and the malnourished and the re-emergence of infectious diseases are, among others, a warning that progress achieved so far towards food security and security in health may be wasted unless effective development policies are formulated and commitments are made to implement them nationally and internationally.

It is abundantly clear that the path to achieving long-lasting food and health security is sustainable development, and sustainable development includes the management and conservation of the natural resource base. Amongst the natural resources that are indispensable for human welfare and socioeconomic development, water ranks as number one. However, scarcity and misuse of this life-supporting resource pose a serious and growing threat to food security, human health and the environment. The world has witnessed the benefits and the disadvantages that have resulted from the wise use and misuse of water respectively. Improper water resources development, based on inappropriate water policies, will have to be replaced, without further delay, by integrated and conservation oriented water resources development policies and programmes.

It is abundantly clear that if the world as a whole and developing nations in particular have to bridge the 'food production and consumption gap', greater growth in crop production is needed. In this context, the importance of irrigation needs no emphasis. While there is no doubt that irrigated agriculture will continue to dominate the food production scene, there is need to incorporate measures to protect human health and the environment in existing irrigation projects and in all new irrigation developments.

The need to incorporate measures to mitigate adverse health impacts with special reference to the control of vector-borne diseases in agricultural development projects gave birth to the WHO/FAO/UNEP/UNCHS Panel of Experts on Environmental Management for Vector Control (PEEM) in 1981. The objective of PEEM is to create an institutional framework for effective interagency and intersectoral collaboration by bringing together various organizations and institutions involved in agriculture, health and the protection of the environment, with a view to promoting the extended use of environmental management measures for vector control within health programmes and in development projects. Since its establishment PEEM has contributed significantly to the application of environmental management measures to control vector-borne diseases in agricultural developments in many developing countries. PEEM achieved this by adopting a comprehensive programme focusing on awareness creation, knowledge synthesis, research and development and capacity building.

Water Resources: Health, Environment and Development, containing the proceedings of a PEEM sponsored international conference held in Brisbane, Australia, is dedicated to help developing countries to adopt environmental management measures to control vector-borne diseases in water resources and agricultural development projects. The proceedings cover a number of critical and relevant issues related to water resources development in the context of food security, health and environment, notably, policy perspectives, multi-disciplinary research, health opportunities in water resources developments, environmental health monitors and indicators in fresh water systems, and socio-economic and health impacts of water resources developments. This book is a welcome addition to PEEM's list of publications and to the global literature on agricultural, health and environmental implications of water resources developments. The task of adopting policies and formulating and implementing programmes conducive to sustainable food production to eradicate poverty, hunger and malnutrition is a major challenge of the twenty first century. What is even more challenging is achieving food security while safeguarding human health and protecting the environment. I have no doubt that this publication will contribute to meeting these challenges by the world community in general and the developing country nations in particular.

Professor A. Sawadogo
Assistant Director-General,
Food and Agriculture Organization of the United Nations
Rome, Italy, October 1998

Preface

I have always been inspired by the book edited by the late Neville Stanley and Michael Alpers, *Man-made Lakes and Human Health* (Stanley and Alpers 1975). In fact, as a young man I had great admiration for the indefatigable vigour and entrepreneurial style of Neville Stanley who, as the leader of an arbovirus group at the University of Western Australia, first established the Australian connection between the academic pursuit of studying viruses carried by mosquitoes and other blood-sucking arthropods (ticks, mites, blackfly, biting midges and sandflies) with the real world of water resource development. His programme has survived him as can be seen in Chapter 8, 'Ord River irrigation area: the effect of dam construction and irrigation on the incidence of Murray Valley encephalitis virus' by John Mackenzie and Annette Broom.

Water Resources: Health, Environment and Development is designed as a follow-up to this original publication and aims to provide readable information to non-specialists, especially those involved in design, planning, construction and management of water resource developments. It would not have been possible without the commitment and foresight of the Townsville-Thuringowa Water Supply Board in northern Queensland, ably supported by the Land and Water Resources Research and Development Corporation, Canberra.

Because there are several good general texts available, this book is not comprehensive and deals primarily with categories 3 and 4 of the communicable diseases associated with water, i.e. water-based diseases and water-related diseases as follows:

- Water-based diseases

  Water provides the habitat for intermediate host organisms in which some parasites pass part of their life cycle. These parasites are later the cause of helminthic diseases in people as their infective larval forms in fresh water find their way back to humans by boring through wet skin (schistosomiasis), being ingested with water plants, crustaceans, or fish that are eaten raw or inadequately cooked (liver and lung flukes), or being infected by a minute crustacean (Cyclops water fleas) and being swallowed (dracunculiasis).

- Water-related diseases

  Water may provide a habitat for water-related insect vectors of disease. Mosquitoes breed in water and the adult mosquitos may transmit malaria, filariasis, and virus infections such as dengue, yellow fever, and Japanese encephalitis. Different mosquitoes vary in their preference for different water bodies but are usually very specific in their requirements. Most malaria vectors require relatively clean water. The culicine mosquitoes, which spread filariasis, prefer to breed in flooded pit latrines and other highly polluted water. *Simulium* blackflies, which spread river blindness, breed in moving water, and *Chrysops* deerflies, conveying eye worm (*Loa loa*), prefer muddy swamps. The tsetse flies that spread African trypanosomiasis (sleeping sickness), although breeding on land, bite near watercourses and can be effectively controlled by clearance of the woodland that fringes the water for a distance of a few metres.

  (World Health Organization 1992: 115–116)

Most of the diseases associated with water are communicable and preventable. Fresh water is often considered to be a renewable resource and limitless, but in fact it is not. In many parts of the world, a shortage of fresh water forms a major barrier to agricultural and industrial production and consequently to economic growth. Nearly half of the global population faces risk through contaminated drinking water, inadequate sanitation or inadequate drainage. Since 1950, the area under irrigation has tripled, and although this may imply great food self-sufficiency, it may also imply an enhanced risk of infection. Although water is the key to life, poor water quality due to sewage, industrial and agricultural effluent can mean increased exposure to nitrosamines and other carcinogenic compounds, insecticides such as DDT, and heavy metals. It may also mean infection with a range of enteric pathogens causing diarrhoeal diseases estimated to be responsible for 4 million child deaths per year.

Inadequate sanitation implies reduced levels of personal and domestic hygiene, which lead in particular to the faecal–oral transmission of diarrhoeas, dysenteries, gastroenteritis, etc. but also of various skin diseases such as otitis externa, scabies, louse-borne fever, typhus and pediculosis (head lice). These are well covered by Cairncross and Feachem (1983), McJunkin (1983) and others. In terms of vector-borne disease, mosquitoes, particularly *Culex quinquefasciatus*, breed readily in the polluted waters of pit latrines and open sumps and drains. The World Health Organization (WHO) estimates that 90 million people are currently infected with lymphatic filariasis which, after approximately 15 years of accumulated infection, may lead to elephantiasis.

The list of diseases associated with water is impressive (Table 1) and it should serve to remind all involved with the water industry that there are often biological ramifications of design. It is hoped, therefore, that this book fosters greater intersectoral thought, which in turn may lead to a reduction in the creation of

Table 1 Examples of water-associated infections by category

Category	*Classification*	*Reason for infection*
Water-borne/washed		
Cholera (*Vibrio*)	Bacteria	Lack of personal and domestic
Diarrhoea (*Salmonella, Shigella,*	Bacteria	hygiene due to inadequate
Campylobacter, Escherichia	Bacteria	water and/or sanitation
Rotavirus	Virus	
Amoeba, Giardia)	Parasite	
Enteric fevers (typhoid, paratyphoid)	Bacteria	Lack of personal and domestic
Poliomyelitis	Virus	hygiene due to inadequate
Ascaris	Roundworm	water and/or sanitation
Leptospirosis (Weil's disease)	*Rickettsia*	
Trichuriasis	Whipworm	
Water-washed		
Trachoma	*Chlamydia*	Lack of water
Relapsing fever	Spirochete	Body lice
Typhus	*Rickettsia*	Body lice
Water-based		
Schistosomiasis	Liver fluke	Snails
Dracunculiasis	Guinea worm	Ingested copepods
Water-related		
Malaria	Protozoan	Mosquitoes
Japanese encephalitis	Arbovirus	Mosquitoes
Dengue fever	Arbovirus	Mosquitoes
Yellow fever	Arbovirus	Mosquitoes
Ross River virus	Arbovirus	Mosquitoes
Oncocerca	Filarial parasite	Mosquitoes
Lymphatic filariasis	Filarial parasite	Mosquitoes
Leishmaniasis	Protozoan	Sandflies

new problems. Such thought should not simply be restricted to the immediate area of activity but also to downstream effects and those pertaining to emerging communities. New settlements on man-made lakes often prosper due to improved fishing resources such as occurred at Lake Nasser when the Aswan dam was built. Because of the increased contact of the population with water and the absence of adequate sanitation, urinary schistosomiasis increased tenfold (Kay 1990). Drastic reductions in silt carried by the annual floods negated natural fertilizer accumulation in the downstream agricultural lands and destroyed a sardine fishing industry supporting 150,000 people at the Nile delta. Downstream, subsoil waters became more saline.

This book would not have eventuated without the help of my colleagues, especially those from the Joint WHO/FAO/UNEP/UNCHS Panel of Experts on

Environmental Management for Vector Control (PEEM). This panel was established in 1981 to foster effective interagency and intersectoral collaboration with respect to health, water and land development and the protection of the environment. Titles in *PEEM Guidelines* series (Birley 1989; Phillips *et al.* 1993; Tiffin 1989) provide valuable knowledge on incorporation of health safeguards, on forecasting vector-borne disease implications and on cost-effectiveness. *Health and Irrigation* (Oomen *et al.* 1990) and *Parasitic Diseases in Water Resources Development* (Hunter *et al.* 1993) are useful complementary texts on the subject.

Water Resources: Health, Environment and Development is designed to acquaint the reader with a variety of topics without going into too much detail. Chapters 1 and 2 describe the basic processes occurring in reservoirs and how 'healthy' water resources can be monitored. Health, in Richard Pearson's context means the physical, chemical and biological quality of water in natural and man-made systems.

Robert Bos (Chapter 3) and Arum Kandiah (Chapter 4) present policy perspectives on water resource development through the eyes of, respectively, WHO and Food and Agriculture Organization experts. One can ask: 'How well have we done in developing intersectoral policies and programmes since the advent of *Our Common Future* (WCED 1987) and subsequent tomes?'

Christine Ewan and colleagues have been heavily involved in determining Australian environmental and health impact assessment principles (Chapter 5) and provide a list of the essential steps in developing a workable process. Martin Birley, leader of the Health Impact Programme, Liverpool School of Tropical Medicine (Chapter 6), follows on in this vein and provides details of case studies from West Africa and South-East Asia.

Once we have established the policies and principles, how do we put them into place in the operational context? How do we teach our future leaders? Peter Furu, Danish Bilharzia Laboratory and British and WHO colleagues (Chapter 7) outline the problem-based learning approach to promote intersectoral collaboration.

Chapters 8–10 are Australian: the Ord River irrigation area thirty plus years later, which introduces an interesting concept of possible seeding of regional areas in western Australia with Murray Valley encephalitis virus from a man-made source. Chapter 9 deals with identification of potential vector- and snail-borne disease problems at the Ross River dam in northern Queensland and our attempts to define management solutions that are applicable to tropical Australia at least. In Chapter 10, Richard Russell introduces the issue of creating artificial wetlands for polishing water and the likely problems associated with poor design. Water reuse is an emerging topic, and with it we feel confident that it will introduce mosquito-borne disease and pest problems. As with water resources development schemes involving dams, it is imperative that potential problems with artificial wetlands are solved at the planning and design phase.

Much has been written about the Tennessee Valley Authority (TVA), established in 1933. I would suggest that this authority is the prototype for dealing with water resource issues. In itself, TVA has been a wonderful success story and we are privileged to have Chapters 11 and 12 dealing with diverse aspects of its development and policies.

Unfortunately, as outlined in Chapter 3, vector control has been the centre of a conflict with recreational usage. Aquaculture is big business around the world. In Chapter 13, Bart Snellen examines the development of fish aquaculture along the northern coast of Java. There is a global message here: successful sustainable management is possible, but if neglected vector-borne disease is likely.

In Chapter 14 we examine the socioeconomic aspects of water resource development in Thailand, especially in the Pak Mun project in the north-east. Successful completion of the project has been hampered by problems of resettlement, dynamited rocks landing on houses and loss of fisheries resulting in compensation claims. Finally, in Chapter 15 I examine end point user problems in urban situations, especially with respect to dengue viruses, and explain some of the new intersectoral strategies used to reduce *Aedes aegypti*.

It is hoped that you, the reader, will expand your views on the scope of issues involved in water resources development. I hope that if each one of you takes away one point that you didn't know before, the world will be a better place.

Brian Kay
Brisbane, Australia

References

Birley, M.H. (1989) *Guidelines for forecasting the vector-borne disease implications of water resources development.* Joint WHO/FAO/UNEP Panel of Experts on Environmental Management for Vector Control, WHO/VBC/89.6, Geneva.

Cairncross, S. and Feachem, R.G. (1983) *Environmental Health Engineering in the Tropics – an Introductory Text.* John Wiley and Sons, Chichester.

Hunter, J.M., Rey, L., Chu, K.Y., Adekoli-John, E.O. and Mott, K.E. (1993) *Parasitic Diseases in Water Resources Development.* WHO, Geneva.

Kay, B.H. (1990) Vector-borne disease implications and their control. *Waterlines*, **9**, 3–6.

McJunkin, F.E. (1983) *Water and Human Health.* US Agency of International Development, Washington DC, USA.

Oomen, J.M.V., de Wolf, J. and Jobin, W.R. (1990) *Health and Irrigation.* Vol. 1 and 2, International Institute for Land Reclamation and Improvement, Wageningen, The Netherlands.

Phillips, M., Mills, A. and Dye, C. (1993) *Guidelines for cost-effectiveness analysis for vector control.* Joint WHO/FAO/UNEP/UNCHS Panel of Experts on Environmental Management for Vector Control, WHO/CWS/93.4, Geneva.

Stanley, N.F. and Alpers, M.P. (1975) *Man-made Lakes and Human Health.* Academic Press, London.

Tiffin, M. (1989) *Guidelines for the incorporation of health safeguards into irrigation projects through intersectoral collaboration.* Joint WHO/FAO/UNEP Panel of Experts on Environmental Management for Vector Control, WHO/VBC 89.5, Geneva.

World Commission on Environment and Development (1987) *Our Common Future*. Oxford University Press, Oxford.

World Health Organization (1992) *Our Planet, our Health*. Report of the WHO Commission on Health and Environment. WHO, Geneva, pp. 115–116.

Chapter 1

Reservoirs: environmental process, management and policy

Walter Rast and Jeffrey A. Thornton

1.1 Introduction

Reservoirs, or artificial lakes, are a group of waterbodies that share many of the same characteristics as natural lakes yet differ from them in several important ways, not the least of which include aspects of their environmental processes and their potential to be managed. Reservoirs are located in both water-poor and water-rich regions. In the former, reservoirs are a practical way of storing surface water during times of plenty for later use during times of scarcity. In the latter, reservoirs act as storage areas to slow the passage of water downstream, reduce flood heights, and protect people and property. Many reservoirs fulfil multiple roles, such as hydropower generation, water supply for industrial, agricultural and domestic consumption, and fisheries production, and many are located in relatively close proximity to human settlements. In every case, reservoirs fulfil a critical human need for fresh water, and, in so doing, ensure their continued place in the spectrum of water resources.

1.2 Types and uses of reservoirs

The term reservoir includes many different types of constructed lakes and storage facilities. In this chapter, the term is used to describe bodies of water created by the construction of a dam or other structure specifically designed to create a pool or reservoir of surface water for any number of human purposes. These waterbodies include river-run lakes, created by constructing a barrier perpendicular to the flow of a river, and off-river storages, created by building an enclosure parallel to a stream and supplied with water by either gravity flow or pumping from the river. Figure 1.1 illustrates these basic reservoir types in a schematic way. Variants of these basic forms of reservoirs include cascades, which consist of a series of reservoirs along a single watercourse, and interbasin transfer schemes, which are designed to move water through a series of reservoirs, tunnels and canals from one watershed to another. These latter systems become especially significant in supplying water to major population centres in water-poor areas, such as south-eastern Australia, south-western USA, north-eastern South Africa, and northern Namibia.

The principal functions of reservoirs have historically been management of water quantity, such as in water supply reservoirs and flood control reservoirs, and power generation. Reservoirs have been used in various forms to train (and drain) watercourses and riverine floodplains to permit agricultural development and promote public safety since the earliest days of human civilization. In this regard, William Dugdale, the eighteenth century English author of *History of Imbanking and Drainage*, cites the construction of barriers along the River Nile by the ancient Egyptians as the earliest example of such artificial water control structures. Later, the Romans built elaborate water diversion schemes to supply water to their cities, some of which (e.g. the Proserpina Dam and Cornalbo Reservoir near Merida, Spain, built in the second century) still exist. Perhaps the first modern use of dams to store surface water dates from thirteenth century Europe where weirs were constructed to provide an adequate hydraulic head, or elevation, and supply of water to drive waterwheels and supply power to mills, which were often located adjacent to streams for this reason. Subsequently, permanent settlements were often established around these mills, many of which are now long forgotten, although the cities remain. Between the fourteenth and sixteenth centuries, the uses of these constructs expanded to include facilitation of river travel by incorporating dams into canal and river navigation schemes, and flood control schemes, and spawned the first multiple use schemes with the development of artisanal fisheries within the basins created by these control structures. The first pollution problems associated with urban reservoirs were reported in the River Thames in the early seventeenth century.

By the middle of the nineteenth century, reservoirs were a common feature of the European (and North American) landscape, and since then dams (and their associated reservoirs) have featured increasingly in economic development schemes throughout the world. Many of the largest dams were built in developing countries after the Second World War. For example, the ten largest reservoirs,

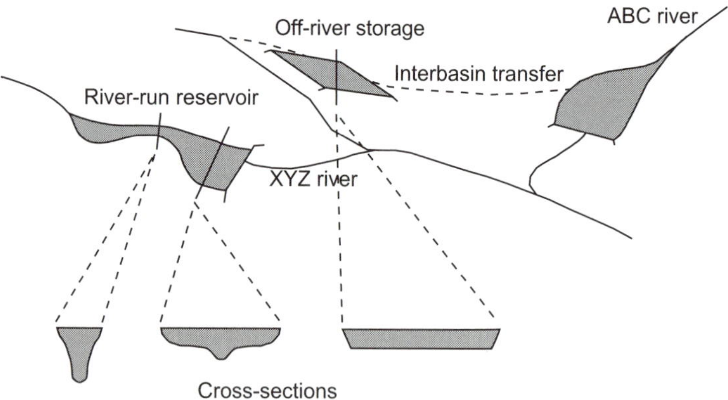

Figure 1.1 Types of reservoirs, illustrating some of their distinguishing characteristics, in a hypothetical watershed.

both in terms of volume of water impounded and height of wall, were all built subsequent to 1955, primarily in developing countries and in countries with economies in transition. Twenty-three countries had more than 100 reservoirs in active operation in 1982, with more than half of the total being located in China, which, in contrast to most other countries in these closing years of the twentieth century, is continuing to pursue a vigorous programme of new construction.

1.3 Characteristics of reservoirs

The characteristics of reservoirs are best described in contrast to natural lakes. One of the first observations that can be made in this regard is that reservoirs are of much more recent origin than natural lakes. As noted above, modern reservoirs date only from about the second century. Their lifespan can be documented in historical time rather than in the geological time-scale applied to most natural lakes. This, together with the fact that they have been created in areas where no natural lakes (or, at least, no significant natural lakes) previously existed, has profound implications for the environment. The exceptions to this general statement are the principally European and North American reservoirs that were created to increase the hydraulic head of natural lakes for the purposes of hydropower generation or river navigation. The consequences of such construction range from social impacts associated with the relocation of people to climatic impacts associated with the enhanced presence of water.

In contrast to natural lakes, reservoirs are often located at the terminal end of their watersheds rather than in the central part of their drainage basins. In addition, the drainage areas of reservoirs often tend to be significantly larger than the drainage areas of natural lakes. This feature ensures that reservoirs are generally subjected to greater contaminant loads than natural lakes. Although the mass of a given pollutant per unit area of watershed is usually similar in both cases, the greater watershed areas of the reservoirs will contribute a greater mass of the pollutant to the waterbodies. This feature not only explains the greater degree of enrichment experienced in many reservoirs, but also contributes to the apparent lack of biological response to such loadings: reservoirs tend to have greater water inflows than natural lakes and thus shorter water residence times, which flush the systems more thoroughly and limit the extent of, for example, algal growth. This is especially noticeable in reservoirs that have water residence times of less than two weeks, which is the typical period during which algal populations develop.

Less obvious characteristics include the gradation of water quality from the upper reaches of a reservoir to the lower reaches, or portion closest to the dam wall. This is especially pronounced in long sinuous reservoirs. A typical gradation in reservoir water quality is illustrated in Figure 1.2. This is in contrast to natural lakes which generally have a more rounded aspect and, hence, a more consistent water quality throughout. Such characteristics in reservoirs can be further modified by reservoir operations. For example, because reservoirs are

– Narrow, channelized basin	– Broader, deeper basin	– Broad, deep, lake-like basin
– Relatively high flow	– Reduced flow	– Little flow
– High suspended solids; low light availability at depth	– Reduced suspended solids; light availability at depth	– Relatively clear; light more available at depth
– Nutrient supply by advection; relatively high nutrient levels	– Advective nutrient supply reduced	– Nutrient supply by interval recycling; relatively low nutrient levels
– Light-limited primary productivity	– Primary productivity relatively high	– Nutrient-limited primary productivity
– Cell losses primarily by sedimentation	– Cell losses by sedimentation and grazing	– Cell losses primarily by grazing
– Organic matter supply primarily allochthonous	– Intermediate	– Organic matter supply primarily autochthonous
– More eutrophic	– Intermediate	– More oligotrophic

Figure 1.2 A typical trophic, or water quality, gradient observed in reservoirs (redrawn from Ryding and Rast 1989).

constructed for specific purposes, provisions are generally made in their design for the withdrawal of water from the basin. In many cases, these draw-off points are located at or near the dam wall (for reasons of ease of construction etc.). Depending on the purpose of the reservoir, withdrawal of water has traditionally been from either the surface (if the purpose of the reservoir was to provide an added hydraulic head for hydropower purposes) or bottom (in most other cases) of the reservoir. This is in contrast to natural lakes where surface water overflow is the most common outlet for lake waters, and can significantly modify not only the thermal and chemical characteristics of the reservoir lake but also have severe repercussions downstream. Withdrawal of water from the surface of the reservoir, shown schematically in Figure 1.3 by arrow A, would result in the discharge of oxygenated, relatively warm water, rich in aquatic life. Withdrawal of water from the bottom of the reservoir (arrow B) would result in the discharge of deoxygenated, relatively cold water. Withdrawal of water from either source generally results in the discharge of water that has a lower suspended solids content than is typically present in the river, due to deposition in the delta (shown in Figure 1.3), leading to enhanced scouring of stream sediments downstream. In contrast, the withdrawal of water from the very lowest portion of the reservoir (through the outlet commonly known as the 'scour' valve, arrow C) can result in the release of substantial amounts of accumulated sediment, contaminants and oxygen-consuming substances into the downstream portions of the watershed with equally severe consequences.

Figure 1.3 also illustrates a further characteristic of reservoirs, which is that the deepest portions of these artificial waterbodies tend to be located at one end of the system, in contrast to natural lakes where the deepest portions tend to be more centrally located (this characteristic is analogous to the locations of the waterbodies within their watersheds). The thermal stratification shown in Figure 1.3 is a common feature between lakes and reservoirs, but even that can be manipulated within reservoirs as a result of reservoir operations. When a reservoir is at full supply level, or its design depth, it may behave in a manner similar to a natural lake of the same size, mixing completely from top to bottom once (in the tropics) or twice (in the temperate zone) per year (in other words, the thermocline shown in Figure 1.3 is present for only part of the year, generally during summer). However, if the pool level, or water elevation, is reduced during the period(s) of annual stratification, the lake may become shallow enough to mix more frequently, resulting in a relatively quick change in water temperature and oxygen levels that can affect both aquatic life and water usage.

As noted, reservoirs are generally constructed for some purpose, often referred to as a beneficial use. These uses generally have associated water quality requirements. For example, water to be used for domestic consumption should have as low an algal content as possible and be free of contaminants injurious to the public health. This requirement will typically necessitate withdrawal of water from a depth below the surface, where algal growth is concentrated, but above the thermocline, below which contaminants such as iron and manganese (which impart an undesirable brownish colouration to the final water) occur. Similarly, water withdrawn for hydroelectric generating purposes should preferably be free of large objects, such as aquatic plants or debris that have accumulated behind the dam wall, and sufficiently well oxygenated (but not too well oxygenated) to prevent pitting of turbine blades. It is not unusual for the beneficial uses of a reservoir to change or be augmented during the life of the waterbody.

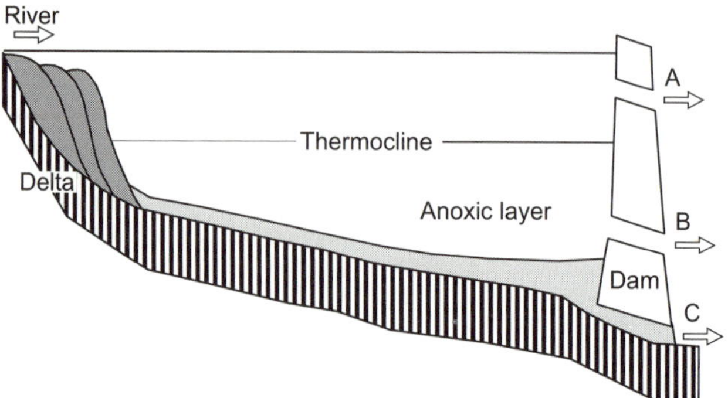

Figure 1.3 Diagrammatic representation of typical points of water withdrawal within a reservoir (redrawn from Chapman 1992).

1.4 Consequences of reservoirs

Thus far, this discussion has been confined to the definition of reservoirs and their distinguishing characteristics, i.e. those features that make reservoirs a distinct type of waterbody. Many of these characteristics have implications for the environment both within the reservoir and elsewhere in the watershed, both upstream and downstream of the impoundment. Some of these implications are a consequence of the age of the reservoir. As noted, reservoirs are constructed lakes having a predetermined life, usually about thirty to fifty years, but some have shorter lifetimes (for example, one South African reservoir was completely engulfed by sediment eroded from within its watershed within 10 years of the date of closure of the wall, although, the benefits of sediment retention were such that construction of the lake was economical), and some longer ones (many European and North American reservoirs/mill ponds are over a century old, while the Roman reservoirs in Spain, described above, are 1,800 years old).

The ageing process that occurs within reservoirs is both different from and similar to that in natural lakes. The ageing process differs from natural lakes in that it is typically shorter. As mentioned, reservoirs can be documented in historical time rather than geological time and segmented into distinct phases, the duration of which can be modified as part of the construction and operation of the reservoir. Figure 1.4 shows a typical example of the lake ageing process drawn from data on Lake Kariba, extending from Zimbabwe to Zambia. Phase I is the filling phase that follows closure of the dam wall. During this phase, organic matter present in the newly created lake basin rapidly decomposes, resulting in extensive deoxygenation and the release of a surge of nutrients into the overlying water. This results in an initial productivity peak or 'trophic surge'. The duration of this phase can be managed to a greater or lesser degree by clearing or partially clearing the lake basin of such material. For example, the period of 'unbalanced eutrophy' was much shorter for Kainji Lake, Nigeria, which was 70 per cent cleared, than for Lake Kariba, which was only 18 per cent cleared. Phase II and III represent a transition phase that can be described in terms of a vulnerable stability and secured stability. Phase II is distinguished from phase III by the dominance of riverine species, i.e. the lake retains many river-like characteristics, while phase III is dominated by lacustrine species, i.e. the lake takes on lake-like characteristics. Also during these phases, the trophic activity within the lake basin switches from a condition of internal nutrient supply, or from autochthonous sources, to one of external nutrient supply, where the reservoir effectively becomes a lake.

At this point, the lake ageing process mirrors that in natural lakes: over time, materials carried in from the watershed continue to accumulate in the lake basin until the depth of water is decreased by sedimentation, and decomposition processes and internal nutrient cycling again dominate. This process continues until the reservoir basin becomes a wetland, a marsh and, ultimately, a terrestrial system. As an aside, limnological studies of Spirit Lake, the natural lake destroyed and recreated by the Mount St Helens eruption in the USA, suggest that newly created natural lakes may also follow all four of these phases, undergoing an initial trophic surge, followed by a period of colonization and stabilization, and ultimate maturity.

Of particular concern during the initial phases of reservoir creation are the effects of inundation. Much knowledge concerning the impacts of inundation on the terrestrial environment was gathered during 'Operation Noah', mounted as a wildlife rescue during the creation of Lake Kariba. Other rescues (e.g. at Lake Brokopondon, Surinam) have been launched since. These efforts have highlighted the tremendous disruption to animal migration patterns that occurs when a lake is created. In addition, follow-up studies conducted after the rescued wildlife is released suggest that even microscale differences in altitude, vegetation, and

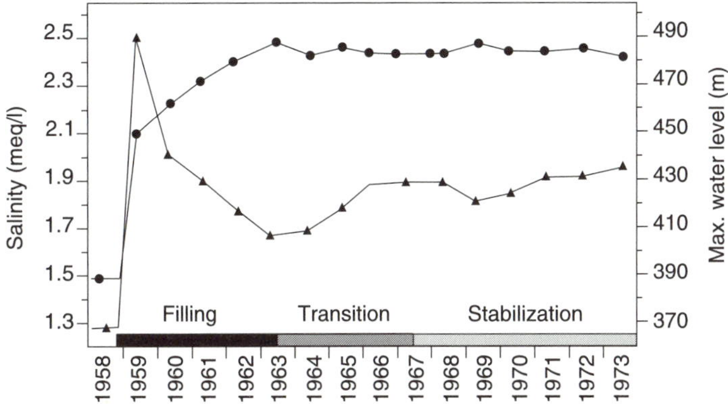

Figure 1.4 The ageing process as observed in reservoirs (redrawn from Allanson et *al.* 1990): salinity (▲); maximum water level (●).

terrain can prove fatal. Little wildlife, in fact, has been successfully relocated as the result of such rescues. The obvious terrestrial impact, therefore, is the loss of endangered and threatened species from within the reservoir basin.

Parallel impacts affect humans and aquatic life. Not only must wildlife be relocated from within the lake basin, but human settlements must also be relocated. At times, many thousands of individuals must be moved, and, while humans can be much more successfully relocated, especially in the developed countries, relocation can cause spiritual and psychological stresses. Moving tribal peoples from their traditional homelands can lead to severe social disruption among primitive societies. Of particular concern is the movement of people to the lake in search of both a stable water supply and economic opportunity, which can 'inundate' the relocated tribal societies and introduce and/or spread water-borne diseases which have a larger-scale impact. Of particular note is the spread of onchocerciasis or river blindness as a result of the construction of Lake Volta, Ghana (Obeng 1975). This lake not only attracted additional people to the lake shore and downstream riparian areas, but modified the river flow regime to provide excellent breeding habitat for the blackfly *Simulium*, resulting in the epidemic spread of this disease. Chapter 14 details problems arising from inundation in Thailand.

Upstream, aquatic life is affected by the creation of the reservoir as the reservoir basin takes on more lake-like characteristics and riverine species are replaced by lacustrine ones. As with terrestrial wildlife, the obvious impact here is also the loss of threatened or endangered species and the restriction of their habitat and pre-impoundment range.

Downstream, the effects of reservoirs generally extend beyond the period of initial stabilization. Some of these have already been noted, such as the epidemic spread of river blindness as a result of the alteration of river flow regimes, the additional potential for downstream discharges to erode the streambed and banks, and the possibility of discharges being sediment-laden or anoxic, which will smother or suffocate river fauna. Numerous additional effects are also possible. Changes in water temperature and/or chemical composition may also affect aquatic flora, while changes in the hydrological regime can adversely affect floodplain flora and fauna, e.g. by modifying the seasonality of flood flows to periods unsuited to the needs of these ecosystems.

In addition, the provision of a 'stable' supply of water may attract ancillary development of both an urban and agricultural nature. Irrigation schemes, popular instruments of national development during the years between the two world wars and in the immediate post-colonial period, imposed an unsuitable crop, usually a monoculture, on soils unsuited for sustained agricultural production. Likewise, urban development, which has traditionally been associated with reservoirs throughout history, imposed, and continues to impose, both demands for water and burdens on sanitation that, through time, often alter the original design purpose of the reservoirs. Land and water use conflicts may be the ultimate consequence of reservoir construction!

The consequences of reservoir construction and operation are numerous. Many such consequences may be unavoidable, e.g. the creation of a reservoir by definition includes creation of a lake. But, in certain parts of the world at least, the presence of reservoirs has mitigated the effects of flood and drought and provided a source of inexpensive power and protein for people. In other parts of the world, the negative consequences of reservoir development have created an anti-impoundment climate, which has, in turn, led to a recognition of environmental needs for water, and a review and refinement of reservoir construction and operation procedures and of water distribution requirements; in effect, a re-evaluation of the necessity for reservoirs in given settings.

1.5 Reservoir management

While the detailed consideration of reservoir construction and operations and water quality management is beyond the scope of this discussion, readers are referred to *Reservoir Limnology: Ecological Perspectives* (Thornton *et al.* 1990) and to *Limnology Now: A Paradigm of Planetary Problems* (Margalef 1994).

Certain aspects of reservoir management should be highlighted: Figure 1.5 presents a decision tree approach to defining, forecasting and mitigating, insofar as is possible, the effects of reservoir development on the environment. And, while particular operating strategies will ultimately depend on the purpose for which the reservoir was created, currently available analytical and management techniques can be used to define the best practicable environmental option (BPEO) for any given reservoir. This option integrates environment and hydrology to identify critical areas within the reservoir watershed, both up- and downstream of the reservoir, and, against this background, seeks to optimize ecological/environmental, social, economic and technical aspects of reservoir operation.

Reservoirs are ideally suited to management. Unlike natural lakes, water resource professionals have the ability to site reservoirs and operate reservoirs in such a way as to minimize their negative impacts on the environment. Modern, multiple level outlet structures, when combined with even elementary lake monitoring programmes, can be used to modify the physical and chemical characteristics of discharge, adjusting depth of abstraction to select for various water qualities depending on intended usage and varying discharge volumes and timings of discharge to retain or pass flood flows etc. For example, cascade reservoir systems can be operated in such a way as to selectively retain contaminants in upstream reservoirs in order to optimize water quality in the downstream reservoirs. In addition, the presence of the reservoir, designed for an assured yield, may reduce some of the hydrological uncertainty previously present in a natural system. However, because of the mere presence of the reservoir in a system where no such waterbody existed previously, it is impossible to escape the fact that the reservoir will have some impact on the watercourse, its flora and fauna and human inhabitants.

1.6 UNEP policy on reservoirs

The United Nations Environment Programme (UNEP) has long advocated a holistic approach to the management of fresh water resources. This comprehensive water resources planning approach to the management of both water quantity and water quality in international river and lake basins was formalized by UNEP as the environmentally sound management of inland waters (EMINWA) process. By definition, the inventory and analysis of fresh water resources, water needs and water management in these international water systems extends also to the national rivers and lakes that form an integral part of international water systems. In effect, the EMINWA process incorporates an explicit recognition of the hydrological cycle (and of the modifying effects of human activities, use and misuse on that hydrological cycle) into a structured systematic approach to data collection and analysis, diagnosis, and corrective action/management planning that can provide a comprehensive and agreed basis for the environmentally sustainable and equitable sharing of finite common natural resources.

Local action at all levels of government is fundamental for the protection of water quality and to ensure the delivery of adequate supplies of fresh water resources, particularly for enhancing socioeconomic development throughout the world. Through EMINWA process-based action plans, local actions can also be co-ordinated and designed to meet the overarching national (and international) goals of protecting environmental quality and the human uses of this fundamental and finite resource in an environmentally sustainable manner.

A sustainable fresh water resources management policy is also based on the premise that national actions within international water systems are enhanced by an agreed multilateral approach. Thus, the planning and implementation of comprehensive action plans developed under the EMINWA process appear most achievable through international agreements. A number of these now exist (e.g. for drainage basins of the Zambesi River, Lake Chad and the Aral Sea) and others are underway. Many of these agreements provide for multilateral commissions and for technical support staff to undertake specific advisory and/or co-ordinating roles pursuant to the international treaties and action plans. This organizational structure is entirely consistent with the UNEP philosophy of local solutions for regional problems, and with the fact that fresh water resources are unequally distributed around the world, both spatially and temporally.

Similarly, UNEP encourages the development of local technical and scientific expertise, and as such encourages technology transfer and technical training rather than the implementation of 'external' solutions. In this manner, UNEP policy is supported by the activities of other United Nations agencies such as the Food and Agriculture Organization of the United Nations (FAO), the United Nations Development Programme (UNDP), the World Health Organization (WHO), the United Nations Educational, Scientific and Cultural Organization (Unesco), and the World Bank.

In applying its EMINWA process, UNEP recognizes that reservoirs will continue to play an important role in supplying the freshwater needs of the

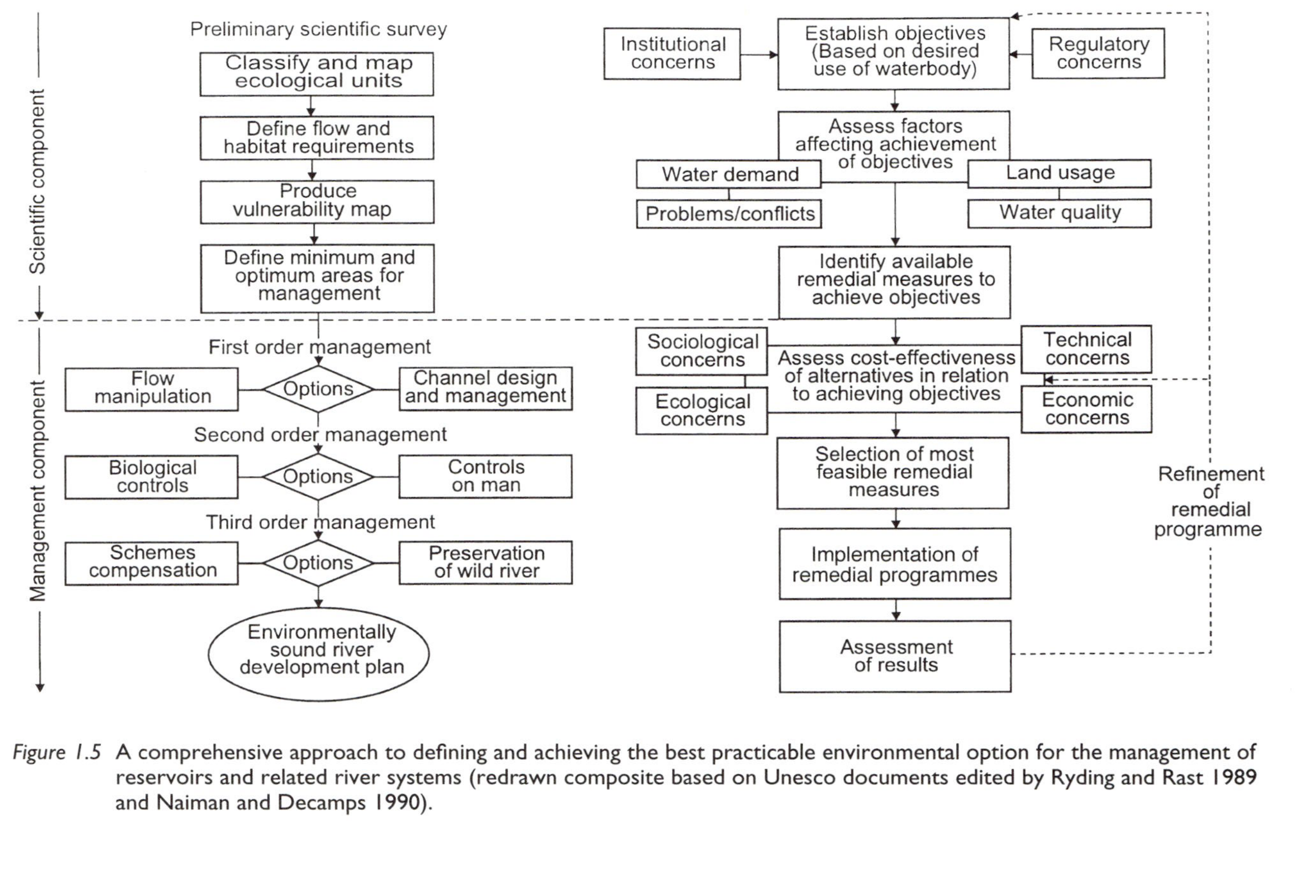

Figure 1.5 A comprehensive approach to defining and achieving the best practicable environmental option for the management of reservoirs and related river systems (redrawn composite based on Unesco documents edited by Ryding and Rast 1989 and Naiman and Decamps 1990).

world's population. This need is nowhere more critical than in those areas where the availability of adequate supplies of fresh water is the primary constraint on economic development. However, the operation of existing reservoirs, as well as the construction of new reservoirs, should be predicated upon a thorough and comprehensive evaluation of the needs of the environment, the needs of people, and the aspirations of the community of nations. UNEP and her sister agencies within the United Nations system stand ready to assist governments in this fundamental process.

References

Allanson, B.R., Hart, R.C., O'Keefe, J.H. and Robarts, R.D. (1990) Inland Waters of Southern Africa. An Ecological Perspective. *Monographiae Biologicae*. Volume 64. Kluwer Academic, Boston, p. 201.

Chapman, D. (ed.) (1992) *Water Quality Assessments: A Guide to the Use of Biota, Sediments and Water in Environmental Monitoring*. Chapman & Hall, London.

Margalef, R. (1994) *Liminology Now: A Paradigm of Planetary Problems*. Elsevier, New York.

Naiman, R.J. and Decamps, H. (1990) The Ecology of Management of Aquatic-Terrestrial Ecotones. *Unesco Man and the Biosphere Series*, Volume 4. Parthenon Publishing Group, Carnforth, p. 316.

Obeng, L.E. (1975) Health problems of the Volta Lake ecosystem, in *Man-made Lakes and Human Health* (eds N.F. Stanley and M.P. Alpers). Academic Press, London, pp. 221–230.

Ryding, S.O. and Rast, W. (eds) (1989) The Control of Eutrophication of Lakes and Reservoirs. *UNESCO Man and the Biosphere Series*, Volume 1. Parthenon Publishing, Carnforth.

Thornton, K.W., Kimmel, B.L. and Payne, F.E. (1990) *Reservoir Limnology: Ecological Perspectives*. John Wiley & Son, New York, p. 246.

Further reading

UNEP (1989) *Sustainable Water Development and Management: A Synthesis*. United Nations Environment Programme, Nairobi, p. 27.

UNESCO/UNEP (1990) The Impact of Large Water Projects on the Environment: Proceedings of an International Symposium convened by Unesco and UNEP and organized in cooperation with IIASA and the IAHS. 21–31 October 1986, United Nations Educational, Scientific and Cultural Organization, Paris, p. 570.

WMO (1992) The Dublin Statement and Report of the Conference: International Conference on Water and the Environment: Development Issues for the 21st Century. 26–31 January 1992, Dublin. World Meteorological Organization, Geneva, p. 55.

Chapter 2

Environmental indicators of healthy water resources

Richard Pearson

2.1 The need for environmental indicators of water quality

The need for environmental indicators of healthy water resources is based on the requirement of human society for a copious supply of water, which is a necessary component of life and of many human activities, including food production, industry and recreation. Given that there is often limited availability of water, careful management of resources is important. This management includes maintaining high quality of such resources as are available, which requires careful monitoring. The purpose of this chapter is to examine briefly the factors that cause water quality to deteriorate, and to summarize some of the methods used to monitor water quality. Some detail of the background to biological monitoring is given so that the appropriate application of biological techniques may be understood.

While the term 'environmental health' is most commonly used in reference to the immediate human environment and its influence on human health, it is also used in consideration of the ecological environment, which may or may not have direct effects on human health. It is in this sense that this chapter addresses the quality of water resources and the means of monitoring this quality. Although the utility of the concept of healthy environments in the ecological sense may be questioned (Calow 1992), the concept does appear to have been adopted generally, and provides a useful context for the non-ecologist to understand natural systems.

The term 'healthy water resources' is defined here to mean the physical, chemical and biological quality of water in natural and man-made systems. While the term 'resource' suggests human utility, it is assumed here that water has worthwhile values whether or not it is used directly by humans. For example, conservation of water bodies for wildlife may have tremendous benefits to plant and animal communities, but may not always have any direct benefit to humans; however, even in these situations there are likely to be indirect benefits to human well-being. Therefore, in this discussion of environmental indicators of healthy water resources, 'environmental health' is used in the sense of environmental quality; it does not include consideration of the water once it reaches the water treatment plant.

This definition therefore looks at waterbodies as providing water for human use, and as a habitat and supply of drinking water for animals and plants. The methodologies discussed here are used to monitor health in this sense. They are limited in that they do not address health in the sense of pathogens: specialized laboratory assessments and water treatments are a normal component of water supply and need not be reiterated here. However, they do provide the means to monitor environmental quality, or health, in the general sense, and in many cases can point to causes of poor quality and environmental degradation.

Monitoring of water quality has traditionally been based on regular sampling of waterbodies and analysis of physical and chemical attributes. More recently, biological methods of assessment have been developed. The discussion below will briefly examine these different approaches. The basic precept of this discussion is that a healthy waterbody (i.e. a waterbody with good water quality) will have sustained chemical and physical quality, reflected in a healthy biota (i.e. the complement of plants and animals). As a basis for monitoring this normally works well, although it will be pointed out that an 'unhealthy' biota, meaning poor water quality, can be normal. It should also be noted that the best quality potable water supply may be unhealthy in the sense that it can provide a breeding ground for disease vectors, such as mosquitoes.

2.2 Water: sources, uses, and management

Water is vital for life, and is one of the basic resources for existence and development of human society. Most of the water on earth resides in the sea, and availability of water on land depends on the hydrological cycle. This cycle involves evaporation of water from the sea to the atmosphere, subsequent precipitation over the sea or land and, on land, transpiration back to the atmosphere through plants, overland flow into streams, flow through the soil, and infiltration and storage or flow in the ground. Water flows back to the sea, along with dissolved and suspended material, via streams, the soil and in the ground. Water for human use is therefore available in surface water (streams, lakes), in ground water and as rainfall. Water is extracted directly from streams and lakes, trapped and stored in dams and weirs, trapped on roofs and stored in tanks, or extracted from the ground via bores and pumps. Sources of water may, therefore, be natural or constructed systems.

Water is required for a variety of human uses, including domestic consumption (drinking, washing, cooking), primary industry (consumption by animals, irrigation, mineral extraction), secondary industry (in industrial processes, for washing, for cooling, etc.) and in tertiary industry (tourist industry etc.). It provides a medium for production of animal and plant species for human consumption, in both natural settings (e.g. fish production from rivers and lakes) and artificial ponds (aquaculture), and habitat for natural communities. Waterways are used for navigation, drainage and waste disposal, and streams and lakes provide major foci for recreation and tourism. In many parts of the world water is in short supply, so these various uses put great pressure on water resources. Therefore, careful management is vital to maintain both quantity and quality of water supply, and to allow for multiple use.

Management of resources includes development of infrastructure such as dams and pipelines, and of delivery strategies. Methods of delivery can have direct bearing on water quality: for example, water extracted from the bottom of dams has high levels of dissolved iron and low levels of dissolved oxygen; and water delivered via stream channels alters the natural flow patterns and degrades the habitat. Dams and extraction from streams can reduce flows, resulting in poor water quality in the stream and diminishing conservation and other amenity values. Management of inflows to natural systems is important in maintaining water quality, and there is much conflict over what should be done about drainage and tailwaters from irrigation areas, mines, industries and other infrastructure such as sewage works.

2.3 Water quality and potential influences on it

Water quality describes the physical, chemical and biological make-up of water. Water is regarded as of high quality when it is safe to drink, that is when its dissolved and suspended constituents are below a level at which they are harmful. However, natural systems may normally experience water quality that is poor by these standards: for example, many parts of the world regularly experience drought during which waterbodies diminish in size and water quality declines; many watercourses drain rocks containing toxic minerals; and the chemical constituents of water differ considerably among geological regions. However, 'poor' water quality is exactly what some organisms are adapted to, and it is at such times that they flourish. The evolution of some scientifically significant species has probably only occurred because of such conditions. It is the species that have evolved tolerance to poor conditions that can provide indication of poor water quality.

Notwithstanding the natural occurrence of water that is unsuitable for human consumption, human influences are responsible for most severe degradation of water supplies worldwide. Given the importance of water as a basic and often scarce resource, contamination of water resources is normally regarded as unwise, but the legislation to prevent it and the wherewithal to police laws and regulations are often lacking, even in first world countries that can well afford it.

Sources of contamination of water supplies are summarized in Tables 2.1 and 2.2. The tables categorize impacts according to their ultimate cause ('pressure'), the manifestation of proximal cause ('effect', i.e. the 'state' of the aquatic environment), and the ultimate effect on the ecosystem, or 'response' of the plants and animals. This loosely follows the familiar 'Pressure–State–Response' model used in reporting the state of the environment (OECD 1993),

Table 2.1 Major causes of degraded water quality

Major cause (pressure)	*Agent causing impact*
Agriculture/aquaculture	
• Land clearing	Erosion → sediments, Weeds
• Removal of riparian vegetation	Erosion → sediments Nutrients etc. (loss of riparian filter) Increase in light Loss of normal organic input (leaf litter, wood) Weeds
• Grazing	Bank erosion → sediments Mobilization of sediments and nutrients Faeces and urine – organics and nutrients
• Irrigation	Sediments Nutrients Pesticides Altered flows
• Cropping	Sediments Nutrients Pesticides
• Horticulture	Sediments Nutrients Pesticides
• Intensive animal production	Sediments Nutrients
Mining	
• Land clearing; overburden removal	Sediments – inert or reactive
• Extraction	Sediments – inert or reactive Leaching of chemicals
• Infrastructure	Sediments
• On-site processing	Chemical disposal/leakage
• Dewatering	Altered flows and water quality
Industry	
• Waste disposal (effluent)	Toxic chemicals Organic materials Sediments
• Waste disposal (aerial)	Toxic gases Fine sediment
• Site activities	Oils, heavy metals, etc. Chemical spills
• Heat	Hot water
Urbanization and infrastructure	
• Land clearing	Sediments, weeds, etc.
• Residences	Fertilizers (nutrients) Pesticides Garden wastes: sediments
• Sewage disposal	Nutrients Organic materials

continued

Table 2.1 continued

Major cause (pressure)	*Agent causing impact*
• Vehicles	Oils, heavy metals, etc.
• Garbage disposal sites	Chemical leachates
• Water resource development	Altered flow regimes
Recreation	
• Swimming	Nutrient and sediment input and mobilization
	Sunscreens
• Boating	Oils, etc.
	Bank erosion leading to sedimentation

except that in that model 'response' refers to human remedial action. It is a convenient method of demonstrating the links between the effects that can be measured and their ultimate causes, and can suggest preventive measures to deal with the ultimate cause or remedial action to deal with the proximal cause. For example, a major change in the faunal composition downstream of effluent from a sugar mill indicated that the effluent was polluting the stream (Pearson and Penridge 1987), and the species making up the fauna indicated that organic inputs were the cause.

In general, impacts on water bodies can be categorized as being derived from 'point' and 'non-point sources'. Point sources are more readily identified, and include industrial effluents, sewage outfalls, road crossings, specific events such as chemical spills, and so on. Non-point sources can be difficult to identify specifically. They include seepage of chemicals from agricultural land, urban run-off, and acid rain. In many cases, however, non-point pollution of one waterway can become an identifiable point source of pollution in the receiving waterbody.

Waterbodies receive water from their catchments, so water quality is a reflection of those catchments. Processes occurring in the catchment affect soil stability and chemistry, and the chemistry of run-off and infiltrating water. Thus, intensive grazing may lead to loss of topsoil and sedimentation of streams; and intensive agriculture using fertilizers may lead to a substantial increase in dissolved nutrients in run-off.

The immediate connection between a waterbody and the surrounding land is the stream bank, or riparian zone. This zone is characterized by vegetation that is different from that in the surrounding land, and stretches a varying distance perpendicular to the water's edge. It is a most important zone for the structure and ecology of the stream. The vegetation helps to stabilize the stream bank, provides shade reducing weed growth in the stream, provides organic input to the stream, slows overland flow and thereby reduces sediment input to the stream, absorbs nutrients from the adjoining land, and provides habitat for a host of animal and plant species, including species with aquatic larvae and terrestrial adults, many species that are terrestrial but feed on aquatic fauna (e.g.

Table 2.2 Effects of impacts on waterbodies and their biota

Agent causing impact	*Effect on waterbodies (state)*	*Biological consequences ('response')*
Flow regulation	Continuous flow in intermittent streams	Reduction in variety of habitat
	Curtailed flow in perennial streams	Reduced water quality
	Major habitat changes	Removal of stimulus for fish spawning
	Reduction in floods	Barrier to migration
Sediments	Smother natural substratum Smother plants and animals Damage respiratory surfaces Reduce light input Altered flow regimes	Reduction in population sizes of most species
Nutrients	Promote growth of algae Promote growth of weeds	Altered flow patterns and sedimentation Change of habitat and in fauna composition
Weeds	Instream weeds change flow regime Riparian weeds alter habitat, change organic input to stream	Major changes to plant and animal communities
Organic materials	Reduce dissolved oxygen Supplement food available	Major changes in faunal composition
Pesticides	Kill or reduce viability of plants and/or animals	Removal of components of flora and/or fauna
Toxic chemical wastes and spills	Kill or reduce viability of plants and/or animals	Removal of components of flora and/or fauna
Acidity	Acidification of water	Loss of intolerant species
Salinity	Increased salt content of water	Loss of intolerant species
Toxic gases	Dissolved in precipitation → acid rain etc.	Removal of components of flora and/or fauna
Oils, heavy metals	Toxic to plants and animals	Removal of components of flora and/or fauna
Hot water	Kills plants and animals	Removal of components of flora and/or fauna
Altered flow regimes	Affects all aspects of water quality	Change in floral and faunal composition Loss of species
Increased light	Promotes growth of algae	Altered flow patterns and sedimentation
	Promotes growth of weeds	Change of habitat and in fauna composition

kingfishers), and species that make use of the special nature of the vegetation of the riparian zone. Removal of the riparian vegetation, which frequently occurs in farming lands, leads to deteriorating water quality and a very different biota from normal.

Developments such as mining may have intense effects on water quality. Whether or not the target mineral is toxic, it is often associated with toxic substances. Thus coal is inert as far as water quality is concerned, but is usually associated with sulphurous minerals that are mobilized during mining, leading to acidification of streams. The mining process usually produces large quantities of waste material, much of which is readily suspended in water to cause sedimentation. Mines may need to be dewatered, a process that brings to the surface large volumes of ground water, which may be of poor quality and which may be disposed of in natural waterways. Mine wastes suspended in water are often retained in tailings dams which may be breached in unusual rainfall events, dumping large quantities of waste into streams. Sediments derived from mines may release contaminants over a long period: that is, their effects last much longer than the event that deposits them.

Although intensive localized activity such as mining can have severe effects on the environment if not carefully controlled, the impacts of such developments are usually much less than those caused by agriculture and grazing. Especially in the tropics, high-intensity stocking on poor soils and in drought conditions leads to severe damage to the soil and to the banks of streams which cattle use as access to water. When rain eventually falls, the soil is unprotected and is washed away, causing sedimentation in streams. Cattle using remnant water-holes for drinking and wallowing cause immense damage to these habitat refuges, especially through input of faeces and urine, and through extensive disturbance of the substratum and riparian zone. Similarly, agricultural landscapes, especially where there is extensive irrigation, lead to broad-scale deterioration of water quality if not managed carefully. The input of irrigation water can lead to raising of the water-table and salinization, and irrigation tailwaters may transport large concentrations of nutrients and pesticides to their receiving waters.

Intensive agriculture (e.g. feedlots, dairies) and aquaculture may produce an effluent that is rich in organic material and nutrients, which cause eutrophication, which is excessive plant growth (both plankton and rooted plants). Eutrophication may involve blue–green and other algae that produce toxins that can kill animals that drink the contaminated water. This process may also occur in water-holes that are frequented by cattle.

Industrial developments can produce a wide variety of waste products which may be discharged deliberately or accidentally into waterways. Deliberate discharges may be licensed, but often the licensing requirements bear little resemblance to the requirements of the ecosystem, and take no account of the cumulative impacts of a number of different discharges on a single stream. Effluents may be toxic (e.g. cyanides from pulp plants), organic, promoting bacterial growth and deoxygenation (e.g. sugar mill discharges), or locally acute but rapidly diluted (e.g. heated effluents from power stations). In all cases dilution reduces the impact.

Urbanization results in replacement of natural vegetation by buildings, roads and gardens. Drainage systems are simplified and rainfall has reduced access to the ground, which is covered by roads, pathways and buildings. Storm water therefore reaches waterways much more quickly than in natural systems, leading to erosion of natural streams. Drainage water carries the effluent of urban society: garden chemicals such as fertilizers and pesticides, excreta of domestic animals, oils and associated chemicals from roads, and general rubbish. Urban areas are served by sewage works which vary in the quality of the effluent they produce, but which may include organic material and nutrients that can cause eutrophication.

2.4 Impacts on flora and fauna

The variety of potential contaminants summarized above not surprisingly has a variety of effects on the biota (i.e. the flora and fauna, or plants and animals):

- the effect of a single contaminant will vary with its concentration (for example, oxygen is vital for human life but is toxic in high concentrations);
- each species has its own physiological requirements and tolerances, so the effects of a particular contaminant will vary from species to species (for example, low dissolved oxygen is tolerated by some insects and not others (Hynes 1970));
- different life stages of a species may react differently to contaminants, such that a concentration that does not affect adults may kill or hamper development of embryos or juveniles, rendering the environment just as uninhabitable in the long term as if the adults were killed (for example, salmon eggs are smothered by sediments that may not affect adults);
- contaminants in combination may have synergistic effects; that is, their combined effect is greater than the sum of their individual effects. Thus a mayfly, an aquatic insect, may be able to cope with low dissolved oxygen, or low current velocities, but not both in combination (Ambuhl 1959).

Notwithstanding these complications, many general effects of contaminants are well known and can be predicted even for species about which we know very little. For example, we can safely predict that a newly discovered species of rat will succumb to strychnine poisoning much like any other rat, although the precise lethal dose may vary from one species to another. This is because the mode of operation of this toxin is to attack the nervous system, which is very similar in all rat species. Likewise, we can predict that sustained low dissolved oxygen levels will kill fish.

Table 2.1 shows that almost any human development can have some impact on aquatic systems and water quality. Such impacts affect the integrity of these systems and diminish their conservation values; and, significantly in the present context, they affect the utility of the water for humans. While it is possible to treat much polluted water to render it potable, or at least useable, such treatment can be expensive. Even pollution can be expensive: the nutrients that reach streams from farmlands come from fertilizers that the farmer has paid for. Efficient and clean use of water depends on an integrated system of supply and waste management, rarely to be found in modern society.

2.5 Methods of assessing impacts and quality – chemical, physical, biological

Water quality monitoring programmes vary in their design with the perceived importance of their purpose. In 'western' society the quality of drinking water leaving water treatment plants is very carefully and frequently monitored because any suggestion of poor quality water reaching the consumer would have unpleasant consequences for those responsible. Monitoring of water quality in streams to check on environmental condition is done less frequently. Typically, water samples are taken at regular intervals (weekly or monthly) and are checked for a variety of parameters, especially nutrients and other dissolved substances. Unfortunately, the dynamics of water chemistry are such that this kind of infrequent sampling can be of little value.

Many physical and chemical parameters change substantially over short periods of time. Dissolved oxygen, for example, can range from 200 per cent saturation in warm weedy water during the day to less than 40 per cent saturation in the same water overnight. Nutrient levels can be difficult to assess because of the dynamic relationships between water, sediments and plants. It is common for an apparently rich system, with lush plant growth, to have very low nutrient levels in the water because most of the nutrients are locked up in the plants. Death of the plants leads to release of the nutrients and may cause algal blooms in the water.

Occasional samples can be very deceptive from day to day. During a survey of polluted streams in Queensland (Pearson and Penridge 1987) it was found that various parameters indicated poor water quality downstream of a sugar mill effluent. However, when by chance sampling was undertaken at a weekend, none of these parameters deviated from normal unpolluted values. This result was simply because the mill did not operate at weekends. 'Spot' measurements are, therefore, subject to very short-term phenomena and have to be taken very carefully. In this case the presence of fish and shrimps at weekends but not during the week was occasionally a good biological indicator of the day-to-day conditions.

Thus, a programme of monitoring of physical and chemical conditions in waterbodies has to be very carefully planned and executed. It is a job for professional practitioners with appropriate training, and it is a job that has to be standardized. Standard methodologies are discussed below. Just as important as the methodology is the interpretation of results, and the canniness of the practitioner. In the example above of low nutrients in the water of a nutrient-rich system, no matter how carefully and accurately the nutrient levels were measured they could not replace a simple note to the effect that there was abundant plant growth at the time of sampling. This qualitative information was vital in the interpretation, and it is exactly the type of observation that could be missed by an inexperienced or poorly trained practitioner.

Physical and chemical monitoring are particularly required when a known contaminant is being targeted. However, knowledge of the physical and chemical behaviour of the contaminant is required to help plan a sampling programme. In many cases, whether the focus is a general regular scan or a targeted substance, sampling is most appropriately achieved by a series of automatic water samplers which can sample at regular intervals of minutes to days, triggered by specified events such as an increase in discharge.

The short-term limitation of much physical and chemical monitoring is now well recognized, and it is commonplace worldwide to include biological monitoring in a sampling programme. Animals and plants, because of their different tolerances and requirements, can provide very good indicators of environmental conditions. This has long been known; a good example is the use of canaries in the past to indicate the presence of dangerous gas build-up in mines: the canaries were particularly sensitive and succumbed before levels reached the dangerous stage (for humans). In the example above of fish entering reaches of a stream at weekends, the fish provided ample evidence that conditions were fine for fish at weekends! However, a problem with this is that absence of fish does not necessarily mean poor conditions: fish are very mobile and may come and go frequently, and their absence on a particular occasion may mean nothing in particular. However, at a larger scale (catchment, subcatchment), absence of an expected fish fauna can be very informative, and may indicate degradation of an entire system.

Some elements of the fauna are less mobile than fish. Their disappearance over a period of time or along a stretch of stream, when they are present at other times and in other comparable stretches, is very suggestive of conditions being inappropriate for them. These components of the fauna include many species of invertebrates such as insects, worms and snails, which like all animals have their individual requirements and tolerances. Being not very mobile, their presence at a site indicates the suitability for them not only of the present conditions, but also of the conditions over a prior period (which will depend on the mobility and life-cycle of each species). That is, the less mobile fauna provide an indication of the lowest level of conditions over the prior period. Thus, in the case of the mill effluents, while fish came in to stream reaches at weekends, during which water chemistry was clearly suitable for them, absence of many insect species during the weekend would be enough to indicate poor conditions over at least the previous few days. Moreover, some species are especially tolerant of poor conditions, and do well in the absence of more sensitive competitors and predators. The presence of these tolerant species, especially in elevated abundance, is further evidence of polluted conditions having prevailed recently and over a long enough period for these species to have established large populations.

Case study: impact of sugar mill effluent on a stream

The investigation by Pearson and Penridge (1987) provides a useful case study of a polluted stream. The stream receives sugar mill effluent, which contains large quantities of organic matter (mainly sugars and alcohols that remain with the waste water during the evaporative concentration of sugar solution) that is fed on by bacteria. The bacteria respire oxygen as they burn up the sugars and alcohols as a source of energy. The bacteria multiply prodigiously in the rich environment and consequently use up much or most of the dissolved oxygen in the water, which then becomes unsuitable for many animal species.

Figure 2.1 shows the effect on the oxygen levels from upstream of the effluent outfall to several kilometres downstream. At the site upstream of the outfall, high levels of oxygen occur, comparable with those measured on other occasions. Downstream of the outfall, the oxygen level declines rapidly, but further downstream recovery begins as oxygen diffuses into the water from the air, and as aquatic plants produce oxygen by photosynthesis. Associated with this clear pattern in oxygen levels is a substantial change in the fauna (Figure 2.1): upstream of the effluent outfall the fauna is characterized by a variety of animals groups (i.e. high diversity), but these rapidly diminish downstream (lower diversity). The diverse fauna is replaced by abundant populations of just a few species of worms and midge larvae (*Chironomus*), which are known globally to be tolerant of low oxygen conditions (which is where they are found naturally). The absence of a diverse fauna, in comparison with a control site, and the presence of animals absent from unimpacted sites are strong evidence of pollution downstream of the outfall. In this case, the level of pollution was clearly tracked by the ratio of *Chironomus* abundance to the abundance of other chironomids (Figure 2.2) – a very straightforward and cost-effective monitoring tool was therefore available, as chironomids are easy to pick from samples, and *Chironomus* is readily distinguished from the other chironomid species.

Thus, monitoring of the fauna at a number of sites provided evidence of poor water quality. In this case, dissolved oxygen measurements demonstrated the cause; however, had dissolved oxygen been measured during a

continued

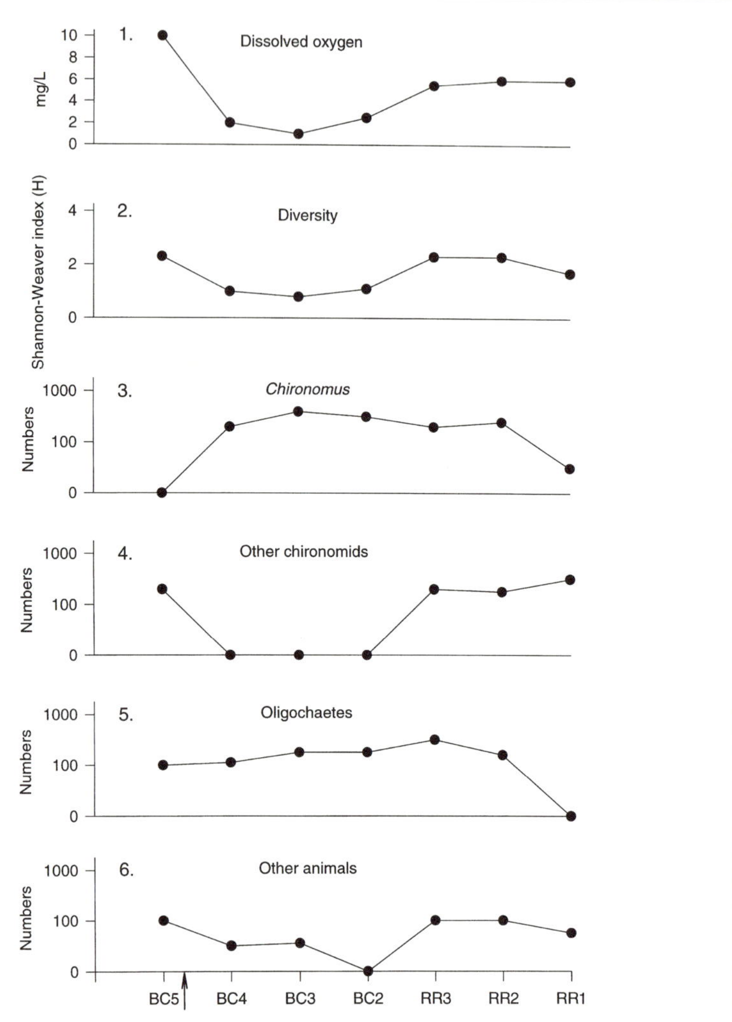

Figure 2.1 Effect of sugar mill effluent on a tropical stream. Sites are arranged from upstream (BC5) to downstream (RR1). The position of the effluent outfall is indicated by the arrow. Commencing from top the figures show: 1 the oxygen sag curve characteristic of organic pollution, caused by bacterial respiration, with recovery taking place downstream; 2 diversity of the invertebrate fauna (Shannon–Weaver index, H) tracking oxygen curve; diversity drops when a few species become dominant, but recovers as more species occur downstream; 3 sudden occurrence of the pollution-tolerant *Chironomus* (a midge larva) with a decline downstream; 4 loss of other, non-tolerant chironomid species; 5 continued presence of oligochaete worms, which are tolerant of organic pollution; 6 loss of other animals in polluted conditions, with recovery downstream.

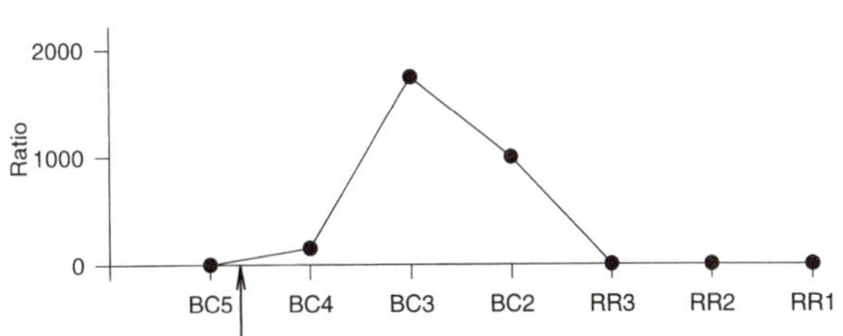

Figure 2.2 Ratio of abundance of *Chironomus* to abundance of other chironomid species. Sites and location of effluent as in Figure 2.1. Note that the ratio is very high in the most polluted sites (nearest the effluent outfall, and indicated by low dissolved oxygen [Figure 2.1]), and is nil or negligible at unaffected or recovery sites.

break in effluent discharge, such as the weekend, no contamination would have been apparent but a 'pollution fauna' would still have been present.

Normally, monitoring designs would aim to include samples before and after the event in question, with impact and control sites, all replicated (a so-called BACI – 'before-after-control-impact' – design) to allow for statistical rigour. Often, however, such a design is impossible, so interpretation of results has to be cautious. Nevertheless, data such as those presented here very strongly indicate impact from the single effluent source; following such indication, the cause can be further investigated.

The above case study is a simple example of how biological monitoring can be applied in the field. The discussion highlights the need to consider temporal and spatial scales – timing and placement of samples is of utmost importance. Biological monitoring is excellent in that animals and plants integrate events over time, and integrate different factors. In some cases, where the animals or plants are well known, the cause of apparent contamination can be strongly suggested by the make-up of the biota. In most cases, however, further physical and chemical investigations may be necessary. This type of biological monitoring of streams, especially using invertebrate animals, is well described in many publications. Comprehensive reviews are provided by Rosenberg and Resh (1993) and Norris *et al.* (1995), multi-author volumes that examine the development and current practice of biological monitoring using stream invertebrates and other plants and animals.

A second form of biological monitoring is bioassay. This usually involves passing the water to be tested through tanks holding animals (usually fish) that have a known tolerance to the contaminants being targeted. The behaviour of the fish in the presence of contaminants is well documented. Fish behaviour is monitored automatically, and when changes take place in normal behaviour, contamination is indicated. This methodology produces immediate results that

Table 2.3 Utility of different indicators at different scales

Indicators	*Spatial scale*	*Temporal scale*
Fish (field)	Catchments	Long term
Fish (field)	Sites	Immediate term
Algae, plants (field)	Sites/reaches/catchments	Short to medium term
Invertebrates (field)	Sites/reaches/catchments	Short to medium term
Fish, invertebrates (bioassay)	Sites	Immediate
Physics, chemistry (water)	Sites/reaches	Immediate
Physics, chemistry (sediments)	Sites/reaches	Medium term

incorporate reactions to any mixture of contaminants. Norris *et al.*(1995) include descriptions of such techniques.

Different methods are clearly applicable at different spatial and temporal scales. Table 2.3 summarizes the applicability of methods to different scales.

2.6 Current methodology

Current methodology includes standard methods of physicochemical sampling and analysis, and various biological techniques, including bioassays, early detection systems, presence/absence of indicator organisms, single population studies, community structure and bioaccumulation (Eaton *et al.* 1995; Humphrey *et al.* 1990) and biological monitoring. This discussion concentrates on bioassay and biological monitoring because of their utility in giving an integrated assessment of the 'health' of the system in question, whereas physical and chemical methods have to be related to known individual standards and species' tolerances to each individual chemical; however, selection of suitable indicators is a crucial process (Cairns and Pratt 1993; Cairns *et al.* 1993). Recent publications that review contemporary methods and provide a good introduction to the relevant literature are *Freshwater Monitoring and Benthic Macroinvertebrates* (Rosenberg and Resh 1993), and the March 1995 issue of the *Australian Journal of Ecology*. Entitled 'Use of biota to assess water quality', this issue comprises a multi-author volume of 19 papers (Norris *et al.* 1995), including discussions of bioassay (Chapman 1995a), ecotoxicity (Chapman 1995b), algae (Whitton and Kelly 1995), diatoms (Reid *et al.* 1995), fish (Harris 1995) and invertebrates (Chessman 1995; Resh *et al.* 1995).

2.6.1 *Bioassay*

Bioassay can be particularly useful in monitoring specific types of conditions. For example, the impact of uranium mining on water quality in the Alligator Rivers region of northern Australia is monitored by passing a continuous flow of stream water through tanks that hold fish. The fish have known responses to the contaminant in question, and monitoring of the fish responses provides a surrogate for monitoring the contaminant itself. Of particular importance is the integrative nature of the fish's response. Other factors (e.g. water temperature, oxygen content) may act synergistically with the contaminant to enhance response; it would be impossible to describe and monitor all potential synergists, so the subject of bioassay provides an effective solution. Cairns and Pratt (1993) and Cairns *et al.* (1993) provide discussion and references on the scope of bioassay methods and their limitations.

2.6.2 Monitoring of natural communities

As discussed above, monitoring of the condition of natural assemblages of plants and animals allows an assessment of the condition of the environment they inhabit. It allows a bottom-up assessment of maintenance of established standards (Cairns *et al.* 1993; Hellawell, 1986). Procedures to be followed in developing a monitoring programme are:

- Define goals:
  - to monitor aquatic system to ensure it is maintained in a healthy state.
- Define objectives:
  - to describe biota (or selected components) regularly
  - to assess against known healthy biota
  - to identify source of contamination if biota below standard
  - to implement remedial action.
- Define methodology:
  - determine frequency of sampling
  - determine sampling methods – what assemblage to be targeted, whether quantitative or qualitative, number of sites, number of replicates, etc.
  - determine level of taxonomic resolution of biota
  - determine analytical methods and level of reporting.
- Formulate action plan:
  - including above considerations.
- Estimate cost of plan:
  - what are the relative costs of different approaches?
  - how do different approaches affect objectives, goal and costs?

The procedures involved in sampling, sorting and identifying the animals that make up the community and that collectively provide the assessment of system health can be expensive if fully quantitative data are required. For example, a quantitative monitoring programme might require ten samples per site on each occasion to provide the required level of statistical discrimination and power.

While each sample would require only a short time in the field, in the laboratory it may take several hours to pick the animals from the extraneous matter (sediments, detritus) and then identify them all. This procedure is undertaken in some places, e.g. in the UK (Wright 1995), but is considered too expensive elsewhere. Reductions in costs are being achieved by the introduction of rapid assessment techniques, which may include non-quantitative sampling, identification of animals to family rather than species level, and development of indices of health on the basis of the presence or absence of various elements of the fauna (Resh and Jackson 1993).

Australia provides an example of the application of rapid assessment that is undertaken as part of a continent-wide 'Monitoring River Health Initiative' (Davies 1994). This involves qualitative sampling of major habitats at each site, sorting either on-site or in the laboratory, and identification to family level. This level of identification reduces some of the power of the technique, but in a continent where the fauna is very incompletely known, it is a pragmatic and cost-effective approach. Analysis uses multivariate models that can be used to predict expected fauna at new sites, and to test whether the new site conforms to the prediction, and is, therefore, healthy or suffering some sort of impact.

Interpreting the results from this sort of programme needs care. Natural processes can have impacts just as severe as those caused by human impact. Recent flooding or drought, for example, may have major effects on the abundance and composition of the fauna. Therefore a variety of information, including climate and discharge records and physicochemical data, as well as the key biological information, is always incorporated into the determination of environmental condition and its causes.

2.7 Developing a regional methodology

Monitoring programmes can be developed at any scale, from catchment to nationwide (Wright 1995) or continent-wide (Davies 1994). However, the more localized a scheme, the greater is the precision that might be built in to it because of the greater concentration of effort. The case study described above allowed development of a straightforward and cost-effective monitoring system for a single river system, which subsequently was found to be applicable to rivers in the region. Developing a monitoring methodology for a waterbody depends on basic data being available or being collected in the initial stages of the programme. While application of a monitoring programme can be undertaken to a great extent by non-specialist personnel, the development phase and ongoing supervision and data interpretation demand the know-how of an expert practitioner. Transfer of general methodology from one place to another is feasible, but for each new region specific information is required, relating to the physical and chemical environment, and especially to the normal regional flora and fauna. While this can be a sticking point, particularly when the biota are not well known, basic ecological surveys can be readily done to describe the biota and set the scene for monitoring. Like the biota of any environment, the flora and fauna of fresh waters varies geographically; nevertheless, there are common components worldwide, especially at higher taxonomic levels, that allow general interpretations of comparisons between sites with even preliminary knowledge of the biota.

Development of an appropriate methodology should be straightforward given the voluminous literature, and should avoid the temptation to reinvent the wheel. Major impediments in the past have been:

- the reliance on physical and chemical methods, which are by themselves inadequate
- ignorance on the part of management personnel of the availability of biological methods
- the absence of suitably qualified staff in management agencies and industries.

These impediments appear to be reducing with more recognition of the importance of the biological environment, and with more general acceptance of the methodology. In particular, recognition at high levels of government has moved on the process of development of appropriate methods and programmes in several countries.

Monitoring is not, of course, sufficient: mechanisms for responding to monitoring information are required. It appears that these too are being improved in first-world nations; unfortunately, however, there is a perception in many parts of the world that environmental health is too expensive to bother about. Our major challenge is to swing opinion to the realization that ignoring environmental health is the more expensive option in terms of human health, productivity and quality of life, as well as in terms of conservation of the natural environment and its plants and animals.

Some of these changes of course can have indirect bearing on the presence of intermediate hosts and vectors of pathogens. The trick is to think broadly about all possible ramifications.

References

Ambuhl, H. (1959) Die Bedeutung der Stromungals ökologischer Factor. *Schweizerische Zeitschrift Hydrologie*, **21**, 133–264.

Cairns, J. Jr. and Pratt, J.R. (1993) A history of biological monitoring using benthic macroinvertebrates, in *Freshwater Monitoring and Benthic Macroinvertebrates* (eds Rosenberg, D.M. and Resh, V.H.), Chapman & Hall, New York and London, pp. 10–27.

Cairns, J. Jr., McCormick, P.V. and Niederlehner, B.R. (1993) A proposed framework for developing indicators of ecosystem health. *Hydrobiologia*, **263**, 1–44.

Calow, P. (1992) Can ecosystems be healthy? Critical consideration of concepts. *J. Aquatic Ecosystem Health*, **1**, 1–6.

Chapman, J.C. (1995b) The role of ecotoxicity testing in assessing water quality. *Australian Journal of Ecology*, **20**, 20–27.

Chapman, P.M. (1995a) Bioassay testing for Australia as part of water quality assessment programmes. *Australian Journal of Ecology*, **20**, 7–19.

Chessman, B.C. (1995) Rapid assessment of rivers using macroinvertebrates: a procedure based on habitat-specific sampling, family level identification and a biotic index. *Australian Journal of Ecology*, **20**, 122–129.

Davies, P.E. (1994) *River Bioassessment Manual, National River Processes and Management Program, Monitoring River Health Initiatives*. Land and Water Resources Research and Development Corporation, Canberra.

Eaton, A.D., Clesceri, L.S. and Greenberg, A.E. (eds) (1995) *Standard Methods for the Examination of Water and Wastewater*. American Public Health Association, Washington, DC.

Harris, J.H. (1995) The use of fish in ecological assessments. *Australian Journal of Ecology*, **20**, 65–80.

Hellawell, J.M. (1986) *Biological Indicators of Freshwater Pollution and Environmental Management*, Elsevier, New York.

Humphrey, C.L., Bishop K.A. and Brown V.M. (1990) Use of biological monitoring in the assessment of effects of mining water on aquatic ecosystems of the Alligator Rivers region, tropical northern Australia. *Environmental Monitoring and Assessment*, **14**, 139–181.

Hynes, H.B.N. (1970) *The Ecology of Running Waters*. Liverpool University Press, Liverpool.

Norris, R.H., Hart, B.T., Finlayson, M. and Norris, K.R. (eds) (1995) Use of biota to assess water quality. *Australian Journal of Ecology*, **20**, 227.

OECD (1993) OECD core set of indicators for environmental performance reviews. Synthesis report by the group on the state of the environment. *Environmental Monographs*, **83**, OECD, Paris.

Pearson, R.G. and L.K. Penridge (1987) The effects of pollution by organic sugar mill effluent on the macro-invertebrates of a stream in tropical Queensland, Australia. *Journal of Environmental Management*, **24**, 205–215.

Reid, M.A., Tibby, J.C., Penny, D. and Gell, P.A. (1995) The use of diatoms to assess past and present water quality. *Australian Journal of Ecology*, **20**, 57–64.

Resh, V.H. and Jackson, J.K. (1993) Contemporary quantitative approaches to biomonitoring using benthic macroinvertebrates, in *Freshwater Monitoring and Benthic Macroinvertebrates* (eds Rosenberg, D.M. and Resh, V.H.), Chapman & Hall, New York and London, pp. 195–233.

Resh, V.H., Norris, R.H. and Barbour, R.T. (1995) Design and implementation of rapid assessment approaches for water resource monitoring using benthic macroinvertebrates. *Australian Journal of Ecology*, **20**, 108–121.

Rosenberg, D.M. and Resh, V.H. (eds) (1993) *Freshwater Monitoring and Benthic Macroinvertebrates*, Chapman & Hall, New York and London, p. 488.

Whitton, B.A. and Kelly, M.G. (1995) Use of algae and other plants for monitoring rivers. *Australian Journal of Ecology*, **20**, 45–56.

Wright, J.F. (1995) Development and use of a system for predicting the macroinvertebrate fauna in flowing waters. *Australian Journal of Ecology*, **20**, 181–197.

Chapter 3

Water resources development and health: the policy perspective

Robert Bos[1]

3.1 Introduction

Some ten years ago, towards the end of the 1980s, the concept of promoting community health status through a process of careful review and adjustment of public development policies gained considerably in popularity. It was a period when intersectoral action for health had taken on important dimensions in the World Health Organization (WHO) under the inspired leadership of the responsible officer in the Division of Strengthening of Health Services, Dr Aleya Hammad El Bindari. The World Commission on Environment and Development (WCED) had just published its report, *Our Common Future* (WCED 1987). In its wake, sustainability was rapidly becoming a new guiding principle in development. The motivation of many workers in the development arena was strengthened by the ideas put forward in the report: no 'after-the-fact' repairs, full integration of environmental considerations into the planning of natural resources development, a focus on multidisciplinarity and intersectoralism, and rational and effective interagency collaboration between United Nations (UN) specialized agencies that for too long had focused on their core mandate without truly linking into the larger development picture.

Almost overnight, environment had become an issue of concern to everybody; it had ceased to be yet another lobby of vested interests; in this case, a group of driven ecologists. Few critical voices were heard. *Our Common Future* (WCED 1987) carried a positive message: continued economic growth was possible as long as governments adhered to the principles of sustainability. The link between poverty and environmental degradation was convincingly explained, turning poverty reduction, that all-time favourite of the development community, into an important environmental goal as well.

If there were any misgivings, they emerged from the international health community. The feeling was that environment seemed to steal the show without

1 Views expressed in this chapter are those of the author alone and do not necessarily reflect the position or policies of the World Health Organization.

THE PEEM NEWSLETTER

WHO FAO UNEP

PANEL OF EXPERTS ON ENVIRONMENTAL MANAGEMENT FOR VECTOR CONTROL (PEEM)

PEEM NEWSLETTER No 21, AUGUST 1988

THE BRUNDTLAND REPORT
A pathway to sustainable development

In 1983, the General Assembly of the United Nations established the World Commission on Environment and Development, with the assignment to develop a global agenda for change — change in long-term environmental strategies; change in the attitudes and perceptions of the international community with respect to the environmental issues at stake in overall development; and change in the relations between countries at different stages of economic and social development in their common concern for the environment in the transition "from one earth to one world".

For three years, the 22 members of the Commission organised various public hearings in all parts of the world, gathering

"Recently I was asked why health was not discussed as one of the major challenges, confronting the world community, in *Our Common Future*. My reply is this: ultimately, the entire report is about health."

Words of Mrs Gro Harlem Brundtland, Prime Minister of Norway, when she addressed the 41st World Health Assembly in May 1988, in her capacity of Chairman of the World Commission on Environment and Development. The conclusions of this Commission are summarised in the following article, and their importance for PEEM is assessed.

agencies (to clean up the mess). . . . But much of their work has of necessity been

ecological systems will not change; the policies and institutions concerned must."

Figure 3.1 The Brundtland Report promoted the idea that global development strategies should consider economics and environment as equal partners, and decried the narrow sectoral outlooks of government ministries.

due attention for the health dimensions. Certainly, *Our Common Future* (WCED 1987) contained a chapter on health, population and nutrition, but only a few specific 'health issues' were mentioned in it. Interestingly, among the first mentioned are those disease problems associated with water: malaria and schistosomiasis associated with irrigation schemes and dams/reservoirs, and diarrhoeal diseases due to lack of access to safe drinking water. In the other chapters, however, reference to health *per se* was made sparingly, if at all. This issue was raised when the Chairman of the WCED, Dr Gro Harlem Brundtland, addressed the 41st World Health Assembly in 1988 (Figure 3.1).

> Recently I was asked why health was not discussed as one of the major challenges, confronting the world community, in *Our Common Future*. My reply is this: ultimately, the entire report is about health.
>
> (WHO 1988a: 1; WHO 1988b: 178–184)

That statement settled the debate: the public health dimension of sustainable development was now clear, and health had made its appearance as a cross-cutting issue, i.e. an issue that cuts across all public sectors and that in principle all sectors have to assume responsibility for. This recognition underscored once again the distinction between health services (for which the health sector has its mandated responsibility) and health status (everyone's responsibility) and opened new avenues for the promotion of health. Obviously, if environmental impact assessment was going to be a central procedure in sustainable development, so should health impact assessment. And if the nature and magnitude of the links between environment, development and health needed further elucidation, multidisciplinary research should be high on the agenda.

Among the many new ideas on public health that flourished, the role of development policies had already loomed on the horizon for some time. There was a great deal of resistance to the promotion of structural adjustment advocated by the World Bank. Long-term goals of sanitizing developing country economies had, in the short term, created havoc in the social sector: social security, public health and education had been badly affected (Caufield 1996). Perhaps the process could be reversed: focused adjustment of development policies could actually be used to promote health status and some of the ill effects of structural adjustment on health services could be turned back by increased attention to preventive safeguards in development projects.

In the remainder of this chapter, the characteristics of policies will first be reviewed, together with their potential to support efforts of intersectoral action for health, in the context of environment and development. Within the narrower confines of water resources development, associated tropical diseases, some of the phenomena and trends in policy development and adjustment over the past decade will be reviewed, with special reference to sectoral government policies and the policies of bilateral agencies.

3.2 Nature and role of development policies

Governments of sovereign states develop and manage the natural resources pertaining to their territory on the basis of national legislation. Legislation provides a rigid framework of laws and regulations to ensure just, equitable and consistent access to natural resources, and their protection from overexploitation and contamination. At the same time, there is sufficient flexibility and room for the political leadership to pursue its economic and social objectives through alternative pathways, and for the public and private sectors to contribute to the process.

The essential role of water in human life is, in this legal connection, expressed in the plethora of water laws and regulations that characterized ancient civilizations as much as they do modern-day society. Water legislation dates back furthest and is most pronounced in parts of the world where it is a scarce resource (e.g. the arid zones of the Middle East) or where survival depends on intensive management of water (e.g. The Netherlands).

Ministers – the political heads of government departments – establish policies, which are decision-making criteria and guidelines aimed to ensure a consistent effort towards attaining the goals a particular government has set itself. The difference between politics and policies is often so narrow that some languages have only one word for both terms (e.g. in Spanish, *política*) and the difference can only be appreciated by interpreting the context of its use. This political dimension of policy reflects the ephemeral nature of the guidance it provides, as opposed to the more enduring (though by no means static) and independent nature of the legal framework.

Policies are the guiding instruments for decision-making procedures that, collectively, make up a strategy. A strategy refers to the allocation of resources towards achieving certain objectives (Mather and Bos 1989). Individual policies can be classified to fit a hierarchical structure. At the national level, macro-economic policies create the umbrella under which other policies take their place. In countries with a centralized planning system, the policies of individual ministries and the development activities they propose are, in theory, tested for their compatibility with the macro-economic policies. This is usually the responsibility of the Ministry of Planning or a National Economic Development Authority. International conventions and treaties, often on trade but also, increasingly, on environmental issues, and conditions attached to external loans from international development banks may limit the room governments have to reassess and adjust their macro-economic policies. This will be reflected in sectoral and natural resource policies further down in the hierarchy. The classification of policies is not just spread horizontally, over different sectors, but also in the vertical dimension of the government structure, i.e. at the provincial and local level.

Crucial policies deal with the allocation of financial or human resources. Other policy statements are more self-serving in nature, intended to appeal to the public and accommodate electoral purposes, but actually saying very little. However, even policies on basic issues of principle are not binding: when different policies on development issues are at odds, as is often the case when it comes to the economic versus the environmental perspective, politics or public opinion take over to reach a solution.

A relevant example of policy conflicts in an industrialized country occurred with respect to the Tennessee Valley Authority (TVA) and health concerns from the past versus recreation of the future. At the ninth meeting of the WHO/FAO/UNEP Panel of Experts on Environmental Management for Vector Control (Geneva, 11–15 September 1989) the representative of the TVA explained how a policy conflict had arisen over the established water management practices in the man-made reservoirs in the watershed. The meeting report includes the following passage.

> As national and regional priorities change, so do policies, and there must therefore always be a provision for their reconsideration and modification. Sometimes, though, the acute problems that led to the original priority setting might have become latent rather than have disappeared completely, and while public awareness and political pressure favour a policy change, the original goals of such policies should not be ignored.
>
> This was well illustrated by the water management policies established

> by the Tennessee Valley Authority (TVA) in the 1930s. The standards of mosquito control maintained by TVA equalled those maintained in privately owned river impoundments under prevailing public health regulations. The measures included the programmed fluctuation of water levels in the reservoirs, a practice that played a key role in reducing *Anopheles* populations and eradicating malaria transmission from the Valley.
>
> New uses of the reservoirs, including recreation and the promotion of nature conservation had led to a conflict of interests. For recreation, stable water levels during summer and early autumn were required; conservation of certain fish species and of waterfowl required higher water levels in spring to promote fish spawning and the rapid growth of aquatic vegetation for the fowl. The changes in water management regimes would without doubt result in increased mosquito populations, yet the potential risk for the reintroduction of vector-borne disease was not appreciated after three generations of malaria free experience.
>
> In recent years much interest has been directed towards the protection and establishment of wetlands, without paying sufficient attention to their mosquito breeding potential. Consequently, TVA has been faced with a conflict of new policy directives concerning wetlands, existing mosquito control policies and state regulations for impounded water. The use of constructed (artificial) wetlands for the treatment of domestic waste water and its processing for reuse is of particular concern, since these could produce large quantities of potential disease vectors and they are often sited close to populated areas.
>
> (WHO 1991: 17–18)

In this case, the gradual phasing out of specific vector control activities in the TVA programme (see Chapters 11 and 12) over the past couple of years indicates that the conflict has been resolved in favour of the pressures from recreational groups and conservationists. In general in these cases, only a thorough, independent assessment can decide whether adherence to existing policies is a genuine issue or a matter of serving vested interests. In this particular case, abolishing the policy of reservoir management for vector control leaves room for other policies, concerned with land use patterns, to ensure that areas of potential risk, such as artificial wetlands, are planned away from areas of human habitation.

3.3 The link between development policies and health

In the concluding chapters of *Our Common Future* (WCED 1987), the WCED reminds its readership that the unprecedented and accelerating change in the different sectoral areas of economic activity represents many formidable challenges to the achievement of sustainable development. And it then puts its finger on the real, fundamental challenge: the systemic character of these changes, which lock together environment and development, different sectors and different countries. It concludes that separate policies and institutions can no longer cope with these interlocked issues.

Recognizing that the real world of interlocked economic and ecological system will not change, the WCED next points out how policy shifts can ensure that decision-making and regulations are in tune with the new reality. Two approaches to environmental policy are indicated. The first, the 'standard agenda', reflects an approach to environmental policy, laws and institutions that focus on environmental effects. The second reflects an approach concentrating on the policies that are the source of those effects.

Keenly aware that health status is a function of environmental conditions (this has recently been presented in quantified terms) WHO set out to elucidate the links between development policies and health. A review of the literature (Cooper-Weil *et al.* 1990) reveals a number of obvious links and also a number of less direct yet substantial impacts on health.

The Cooper-Weil *et al.* (1990) review considers first macro-economic policies, and then, one by one, agricultural, industrial, energy and housing policies. A number of basic issues, common to all of the above, were identified. First, the magnitude and diversity of health hazards and risks associated with development have been underestimated in the past. Numerous obstacles hinder the assessment of these risks inherent to development policies: lack of procedures and institutional arrangements, lack of technical capacity to deal with the assessment and subsequent implementation of preventive measures and a lack of resources to collect the necessary data and to include the necessary design changes in development projects.

Second, development policies have moved many countries into a transition phase between an agriculture-based economy and an industrialized and services-oriented economy. In this transition, the health sector is faced with a double burden of disease: the traditional infectious diseases combined with non-communicable diseases such as cardiovascular ailments or malignant tumours.

Thirdly, macro-economic adjustment policies, designed for long-term benefits, have, in the short term, a detrimental effect on the health status of vulnerable groups. These include the very poor, children, women and some specific occupational groups.

Because of its organization by economic sector, the links between water resources development and human health appear in various contexts in the review. They include the effects of irrigation development (the introduction of irrigated agriculture into new areas, policies that influence operation and maintenance of existing schemes, water management policies in the light of changing land use and cropping patterns, the effect of new varieties and pesticide use in irrigated areas, the latter related to government subsidies for pesticides to keep food prices low, and the lack of integration between sectors dealing with drinking water supply and those dealing with irrigation), dam construction (with a focus on policies guiding dam design, reservoir management and resettlement of affected communities) and urbanization (urban water management, policies on wastewater use and their health implications, the implications of increasing demands for urban drinking water from remote areas). In brief, not only are there a large number of water policies impacting directly or indirectly on health, but they guide decisions in different sectors that may lead to incompatible development activities.

WHO's contribution to the summit in New York 'five years after Rio' in July 1997 takes an innovative perspective on the health and environment links (WHO 1997). It starts off with a chapter on the driving forces behind current health and environment trends: population dynamics, urbanization, poverty and inequity, consumption and production patterns and economic development. In the next chapter, these driving forces are translated into human activities affecting environmental quality: the production of household waste, the development and use of fresh water resources, land use and agricultural development, industrialization and energy generation.

The exposure to risks posed by these activities through their impact on environmental quality is the subject of the subsequent chapter. And finally, prevailing health conditions are put in the relevant environmental context. In its conclusions this report states, *inter alia*:

> Environmental quality is an important direct and indirect determinant of human health. Deteriorating environmental conditions are a major contributory factor to poor health and poor quality of life and hinder sustainable development.
>
> (WHO 1997: 197–198)

Impoverished populations living in rural and peri-urban areas are at greatest risk from degraded environmental conditions. The cumulative effect of inadequate and hazardous shelter, overcrowding, lack of water supply and sanitation, unsafe food, air and water pollution, and high accident rates impact heavily on the health of these vulnerable groups. Poor environmental quality is directly responsible for around 25 per cent of all preventable ill-health in the world today; with diarrhoeal diseases and acute respiratory infections heading the list. Other diseases such as malaria, schistosomiasis, other vector-borne diseases, chronic respiratory diseases and childhood infections are also strongly influenced by adverse environmental conditions, as are injuries.

3.4 Developments in some sectoral policies

Policy adjustment is a continuous process that has been particularly dynamic during the past ten years as a result of the important changes in the geopolitical situation. Some trends are indicated below.

3.4.1 Environmental policies

As might be expected, the evolution of environmental policies, strategies and decision-making procedures has been rapid over the past ten years. In the 1980s, environmental impact assessment (EIA) was usually something imposed by external donor agencies. Today, most countries have environmental legislation in place that provides the framework for proper impact assessment procedures. These procedures are carried out by an Environmental Protection Agency or Council, which by definition is intersectoral in nature and where all relevant ministries participate in the screening, scoping, development of terms of reference and appraisal of impact assessment reports.

Regrettably, the health sector has been rather timid in joining in. As a result, health is still often viewed in a one-dimensional way (i.e. in terms of the provision of health services) rather than as a dimension of the environmental change that is foreseen. Often, this finds its origin in the fact that the health sector simply lacks the capacity to contribute to the intersectoral decision-making that is necessary, for instance at the impact appraisal stage. Training and improved institutional arrangements offer a solution to this problem.

3.4.2 Health sector policies

Trends in health sector policies reflect to a considerable degree the process of downsizing of public sector administrations, which particularly affects the developing countries. In most countries, the process of decentralization has been set in motion, again in the broader context of government reform. Calls for decentralization often go together with an emphasis on increased intersectoral action for health at the district level. The district is, indeed, the natural place for workers from different ministries to meet and where sectoral boundaries naturally disappear. All too often, however, health workers have not been trained to work in a multidisciplinary set-up and there are several accounts where the vertical allocation of resources, strictly within the sectoral boundaries, leaves no room for effective intersectoral collaboration. National policies that enhance intersectoral competition between ministries for, admittedly, scarce resources have to be identified and adjusted to achieve the full benefits of the decentralization process.

Another important development is the introduction of economic considerations in the priority setting of health sector programmes. With the introduction of the concept of Disability Adjusted Life Years (DALYs), featured prominently in the World Development Report (World Bank 1993), ministries of health have a tool to make a more rational choice in resource allocation. This works very well for decision-making within the health sector. For the introduction of health considerations into development projects of other sectors, it seems less suited.

A typical example of health sector policy reform comes from Zambia, where the following was reported at a national seminar on water resources development and vector-borne diseases in 1995.

Health policies

Aims: The national health policies aim at 'providing Zambians with equity of access to cost-effective quality health care as close to the family as possible'. The concept recognizes the fact that improvement of accessibility to health services and reduction of mortality and morbidity is not enough to provide quality health care, but also the improvement of the quality of life of all Zambians in general. Main determinants in this regard include environment, lifestyle (behavioural), socioeconomic, cultural and political factors and health services.

Achieving a healthy society is not a medical issue but rather a political and socioeconomic one. Accordingly, the policies stress the need for intersectoral collaboration, in particular with other ministries. The Ministry of Health, however, shall strive to promote preventive, promotive, curative and rehabilitative services as a major strategy for achieving health for all with the individual taking responsibility for his health.

Among the health goals targeted by the National Health Policy are:

- achieving equity in health opportunities
- increasing the life expectancy of Zambians
- creating environments which support health
- encouraging lifestyles which support health
- providing quality assured health services
- promoting individual and family health through efficiently administered population control activities.

Some of the specific strategies identified as cardinal to the creation of environments supportive to health, include ensuring safe working environments which support health, and ensuring safe physical environments and health supportive habitats.

Structures

In accordance with the recently enacted Health Services Act of 1995, provision and delivery of health services are in the process of being decentralized to the districts under district health boards. All curative services shall be the responsibility of the boards while environmental health services, such as creating environments supportive to health, shall continue to be a delegated responsibility of the local authorities.

The Act has further created the Central Board of Health, which shall be responsible for the day to day management and provision of health services by way of monitoring the operations of the district health boards.

Public health laws relevant to water resources development

> Given that there are several statutes that regulate various aspects of public health, perhaps the most pertinent to water resources development and management are the Public Health Act and the Examination of Mosquitoes Act. In their broad provisions, they each refer to mosquito control in relation to water resources utilization. It should be so managed that it does not encourage mosquito breeding.
>
> (WHO 1995)

3.4.3 *Water policies*

Following the adoption of Agenda 21 (UN 1993), a lot of developments have taken place with respect to national water policies. The general trend is towards integrated development and management of water resources (further details are given in Chapter 4).

3.5 Policies of external support agencies

In most countries where water resources development has a potentially significant impact on the environmental determinants of human health, such development is implemented through a partnership between government and agencies mandated to carry out technical assistance and investment support. In preparation for the ninth meeting of the joint WHO/FAO/UNEP Panel of Experts on Environmental Management for Vector Control (PEEM) in 1989, a review was carried out of the policy perspectives of this partnership. In the context of such partnerships, the impact of policies of external support agencies is predominant in determining the course of events during planning, construction and operation. The development objectives of these agencies, i.e. poverty alleviation and the sustainable use of natural resources, are reflected in their policies, together with political considerations. This does not, however, imply that the true spectrum of health risks and safeguards associated with water resources development is automatically taken into account.

The importance of bilateral agencies may be less than that of multilateral bodies such as the World Bank and the various regional development banks in terms of the volume of financial resources that they make available to the developing world. Yet their role in setting new trends and testing out new mechanisms of delivering aid is not to be neglected. This, and the fact that they directly report to and financially depend on national parliaments, results in a more interesting and diverse mix of policy aspects.

For the 1989 PEEM review, eleven bilaterals were contacted, most of them through personal visits (Mather and Bos 1989). An overview of the findings is given in Table 3.1. At the end of the 1980s, bilaterals were actively reviewing their remit and policies in the light of recommendations made in *Our Common Future* (WCED 1987). There was optimism that the sustainable development concept would set off a new range of opportunities for developing countries to improve the welfare of their people. There were, however, also concerns. The conclusions of the PEEM review's chapter on bilateral agency policies presented the following picture.

> A common feature among many bilateral agencies is the trend to decentralize staff and expenditure from headquarters. This, accompanied by rapid movement of desk officers and technical staff, reduces the awareness of those personnel in the complex environmental and health issues now becoming a part of the agencies' programmes. Some of the agencies have responded to this by preparing 'issues papers' on these subjects for their staff and counterparts.
>
> In the development-oriented projects of most bilateral assistance agencies, increased attention is being devoted to environmental aspects. The same agencies often have projects with a specific health orientation. Yet there is minimal consideration given to the human health dimension of the environment in development activities. This reveals the same dichotomy between health and development that typifies many of the developing country intersectoral deficiencies that have been so widely criticized.
>
> The fundamental themes of sustained development and production and the relief of poverty, which characterize current bilateral assistance policies, have yet to find operational modalities. Programmes and projects with these aims call for environmental, social and health oriented studies which are necessarily long term, requiring monitoring, review and revision over periods which will generally exceed that of donor support. Yet evaluations and project reviews reveal no provision for such continuity under local operation. In fact, a failure of integration within the local institutional system is one of the commonest criticisms. Environmental management for vector control represents an example of this type of activity, and there is a need for measures to solve the problem, which threatens to invalidate many otherwise justifiable initiatives.
>
> The basic aims of most bilateral assistance are the same as those of the multilateral organizations of the United Nations System, with the main operational differences being the selection of regions, countries and sectors of concentration under programmes of the former. Similar problems confront both, in their attempts to integrate environmental and health components in their development-oriented activities.
>
> In most of the bilateral agencies visited, the donor country's commitment and financial contribution to the UN system was stressed. There was therefore considerable interest in how the UN programmes, funds and specialized agencies most closely involved (UNDP, UNEP, UNICEF, WHO, FAO) intended to address these issues of common concern, and the hope expressed that bilateral and multilateral programmes would be compatible and complementary.

Table 3.1 Characteristics of operation of some bilateral assistance agencies

Characteristics of operation	*Belgium ABOS*	*Canada CIDA*	*Canada IDRC*	*Denmark DANIDA*	*Germany GTZ*	*Netherlands DGIS*	*Italy DGCS*	*Japan JICA*	*Sweden SIDA*	*UK ODA*	*USA USAID*
1. Part of OD aid through:											
a) bilateral programme	1	1	1	1	1	1	1	1	1	1	1
b) multilateral programme	1	1	2	1	1	1	1	1	1	1	1
2. Concentration on 'programme countries' for bilateral aid	2	1	2	1	–	1	1	–	1	–	–
3. Existing policies or strategies on:											
• natural resources development	–	1	2	1	–	1	1	–	1	–	1
• environment	–	1	2	1	–	1	2	–	2	1	1
• health	1	–	2	1	–	1	1	–	1	–	1
4. Implementing field programmes on:											
• natural resources development	1	1	2	1	1	1	1	1	1	1	1
• environment	–	1	2	2	–	1	2	–	1	2	1
• health	1	1	2	2	1	1	1	2	2	2	1
5. Agency guidelines for development cover on:											
• environment	–	1	–	1	–	1	–	–	–	1	1
• health	–	–	–	1	–	–	–	–	1	–	
6. Requirement for environmental impact studies	–	1	–	1	1	1	–	–	2	2	1

Table 3.1 continued

Characteristics of operation	Belgium ABOS	Canada CIDA	Canada IDRC	Denmark DANIDA	Germany GTZ	Netherlands DGIS	Italy DGCS	Japan JICA	Sweden SIDA	UK ODA	USA USAID
7. Desk officers located in:											
a) HQ	1	1	1	1	1	1	1	1	1	1	1
b) region or country	–	1	2	1	–	2	1	1	1	2	1
8. Technical support staff											
a) agency direct	1	1	1	1	1	1	1	1	1	1	1
b) subcontracted	1	1	1	1	1	1	1	–	1	1	1
c) host country	–	1	1	1	–	–	1	–	1	1	1
d) home NGOs	1	1	–	1	1	1	1	–	1	2	1
e) other NGOs	–	–	1	2	–	–	2	–	2	–	1
9. Project evaluation process	–	1	1	1	–	–	1	–	1	1	1

OD = overseas development; 1 = a significant issue for the agency; 2 = a minor issue; – = inadequate information; NGOs = non-governmental organizations.

> Before analysing the role of the UN system, it is appropriate to note that the donor countries with bilateral programmes have a voice in the assemblies, conferences, executive boards, councils and sectoral committees responsible for approving the system's programmes. Their influence on issues of particular interest therefore depends on the expression of concern and support by their delegates and representatives attending these meetings.
>
> Environment and health in natural resources development are now the subject of universal concern, unlikely to raise serious controversy. It is in the hands of the donor countries to ensure that their representatives are suitably briefed to press for compatibility and complementarity among bilateral and multilateral programmes aimed at helping developing countries to integrate these concepts within their plans for development.
>
> (Mather and Bos 1989: 50–51)

In retrospect, one can only be amazed at the speed with which the 1989 version was overtaken by events. The fall of communism in eastern Europe and the former Soviet Union initially sparked off a debate on the 'peace dividend' from which public health was certain to benefit. Soon, however, shifts in the global and regional political balance resulted in instability and conflict. Aid money was diverted to peace-keeping operations, and the pressure on international assistance budgets increased as newly independent states in eastern Europe and Central Asia joined the traditional recipient countries.

So what happened to the issues raised in the 1989 review? As a result of public sector reform, the phenomenon of reduced awareness of more complex cross-cutting development issues and a general lack of 'institutional memory' in bilateral agencies almost certainly increased. In most cases bilateral agencies have become fully integrated into the foreign affairs structure. At the same time, integration usually did not lead to an effective restructuring, beneficial to the cause of 'intersectoralism'. In many agencies, attempts to implement a matrix approach that would do justice to the cross-cutting nature of environment and health bore little fruit, especially with respect to the latter. Resource flows proved rigid and intersectoral walls between different technical desks often remained insurmountable, with few exceptions. The conservative role of health sector specialists themselves in slowing down the process of change should not be underestimated.

In preparation for and also following the United Nations Conference on Environment and Development in Rio de Janeiro during 1992, bilateral agencies produced a constant stream of policy and position papers. Generally speaking, the perception of health in the framework of environmental policies remained one-dimensional. The response was usually focused on the provision of new, and the strengthening of existing, health services.

To think globally turned out to be relatively easy. To act locally remains, however, an elusive objective. Ample resources were, for example, made available for activities related to global climate change (admittedly, mainly from multilateral agencies). Precious few resources, on the other hand, were accessible for an issue such as changing land use patterns. Fortunately, the efforts of a few driven individuals ensured that the health aspects of global climate change were put on the international agenda (McMichael *et al.* 1996). Research on the human health dimension of land use change was much harder to 'sell'. The example of the WARDA/PEEM/IDRC consortium research project (WHO 1993) on the association between rice production systems and vector-borne diseases in West Africa, though now under successful implementation, is a case in point.

The proposal was developed as a multidisciplinary research project to address the important issue of malaria and schistosomiasis risks related to a rapid expansion of rice production systems in West Africa. As such, it contained all the relevant research questions to provide the basis for more sustainable agricultural development. The co-ordinating institution, the West Africa Rice Development Association (WARDA), works regionally in 17 countries in West Africa. For the purpose of the project, a consortium was established consisting of WARDA and national health and agricultural research centres in Ivory Coast and Mali. The study sites were selected along a north–south transect across the different ecological zones of West Africa. The expected outputs are to serve as decision-making tools that will allow agricultural ministries in West Africa to select the best options (in terms of health risks) for the expansion of rice cultivation.

In terms of objectives, approaches, partnerships and expected outputs, as well as in spirit, the proposal fitted the criteria laid down in the new policies of the vast majority of bilateral agencies. Its multidisciplinary nature also conformed to the recommendations made by the Commission on Health Research for Development (CHRD 1990). When it came to appraising the proposal, however, most agencies did not have the structure and procedures needed to handle it effectively. Typically, the proposal would be passed from one technical desk to another, and end up in the 'intersectoral gaps' between them since no one was willing to take accountability for it or because core funds could not be used for a cross-cutting project that was marginal from each individual desk officer's viewpoint.

It was fortunate that the project had been developed together with the Canadian IDRC, often listed among bilateral agencies but in reality an institute whose acronym stands for what it is: a centre carrying out international development research. As such, IDRC has not only adjusted its policies to fit the concept of sustainability development, but it has also adopted a revolutionary structure and set of procedures favouring a multidisciplinary focus.

The exceptional progress made by the bilateral agencies of the Nordic countries deserves mentioning here, not because two of them (Danida and the Norwegian government) decided to provide support to the WARDA/PEEM/IDRC consortium research project (WHO 1993), but because they seem to have managed to successfully translate their policy adjustment into structures and procedures. This is demonstrated by their programmes, and with regard to water and health issues has resulted in such initiatives as the Stenungsund Conference sponsored by SAREC/SIDA (SAREC and TDR 1994) and subsequent support for the land environment and tropical diseases component in the UNDP/WB/WHO tropical diseases research programme, the Danida support for the PEEM/DBL training course 'Health Opportunities in Water Resources Development' (Chapter 7) and their growing interest in the area of wetland conservation and human health.

References

Caufield, C. (1996) *Masters of Illusion. The World Bank and the Poverty of Nations.* Macmillan, London and Basingstoke.

Commission on Health Research for Development (1990) *Health Research, Essential Link to Equity in Development.* Oxford University Press, Oxford.

Cooper-Weil, D.E., Alicbusan, A.P., Wilson, J.F., Reich, M.R. and Bradley, D.J. (1990) *The Impact of Development Policies on Health.* World Health Organization, Geneva.

McMichael, A.J., Haines, A., Slooff, R. and Kovats, S. (eds) (1996) *Climate Change and Human Health.* World Health Organization, Geneva.

Mather, T.H. and Bos, R. (1989) Policies and Programmes of Governments, Bilateral and Multilateral Agencies and Development Banks for Environmental Management in the Context of Natural Resources, Agriculture and Health Development. Document WHO/VBC/89.7, World Health Organization, Geneva.

Swedish Agency for Research Cooperation with Developing Countries (SAREC) and TDR (1994) *Parasites of Poverty. Highlights from a seminar on tropical diseases, society and environment.* SAREC/SIDA, Stockholm.

United Nations (1993) *Agenda 21: Programme of Action for Sustainable Development.* UN, New York.

World Bank (1993) *World Development Report 1993: investing in health.* Oxford University Press, New York/Oxford.

World Commission on Environment and Development (1987) *Our Common Future.* Oxford University Press, Oxford.

World Health Organization (1988a) *PEEM Newsletter*, no. 21, Secretariat of the Joint WHO/FAO/UNEP Panel of Experts on Environmental Management for Vector Control, Geneva.

World Health Organization (1988b) Interview with Mrs Brundtland. *World Health Forum* **9** (2): 178–184.

World Health Organization (1991) Report of the ninth and tenth PEEM meetings, with technical discussions sessions on policies and programmes and on livestock management and disease vector control. Document WHO/CWS/91.11, World Health Organization, Geneva.

World Health Organization (1993) WARDA/PEEM initiative for a consortium research project on the association between rice production and vector-borne diseases in West Africa. Unpublished document, PEEM Secretariat, World Health Organization, Geneva.

World Health Organization (1995) *Water resources development and vector-borne diseases in Zambia. Report of a national seminar, Kafue Gorge, 6–10 November 1995.* World Health Organization, Geneva.

World Health Organization (1997) *Health and Environment in Sustainable Development. Five years after the Earth Summit.* WHO/EHG/97.8, World Health Organization, Geneva.

Chapter 4

Water resources development: policy perspectives of the Food and Agricultural Organization in relation to food security

Arumugam Kandiah

4.1 Introduction

Water sustains life; it is indispensable for the welfare of human beings and their natural environment. But scarcity and misuse of this life-supporting resource pose a serious and growing threat to food security, human health and well-being, industrial development and the ecosystems on which they depend. Scarcity or poor quality can have a dramatic impact not only on agricultural production, but on all aspects of everyday life, very often adding to the burden of women. There is an urgent need for judicious management of water resources worldwide.

The world has witnessed the benefits and disadvantages of, respectively, the wise use and misuse of water as well as the hazards posed to natural resources, human health and environment by improper water resources developments based on inappropriate water policies. The Food and Agriculture Organization (FAO) believes that in order to overcome current deficiencies and to meet the total water needs of the human population, the management of water resources must form part of the spectrum of development endeavour. This approach offers the prospect of sustainable use of available water supplies as well as cost-effective solutions to present-day water problems. A wider understanding among different users, where each attaches a different value to it, that misuse by one group can adversely affect the others would be a major step towards achieving effective water management and harmony.

4.2 The hydrological cycle

Fresh water is a renewable resource by virtue of the hydrological cycle, but for all practical purposes it is a finite one. The hydrological cycle is a continuous process by which water is transported from the oceans to the atmosphere to the land and back to the sea. Every year solar energy converts 500,000 km^3 of water from sea and land into water vapour, which eventually returns to the earth's surface as rain and snow (Figure 4.1).

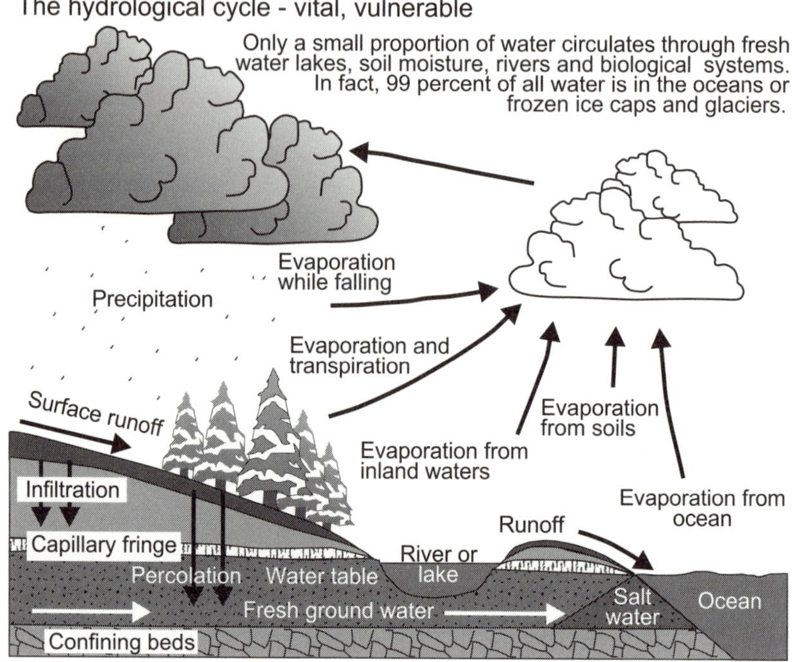

Figure 4.1 A schematic representation of the hydrological cycle.

At any given time, only a small fraction of the world's total water resources is under circulation; a large portion, about 99 per cent, is in the oceans or frozen in the ice caps and glaciers. About 40,000 km^3 runs off the land and into the seas every year, of which only 9,000 km^3 is readily available for human use (Clarke 1991). This is the fraction of the total water under circulation that the world's human population is dependent upon for its fresh water supply. One thing should be clearly understood: for all practical purposes, fresh water should be considered to be a finite resource; there is a fixed amount on the planet that can neither be created nor destroyed.

4.3 Availability of fresh water

Many factors affect the availability of water to humans. Much rainfall, particularly in the tropics, occurs during heavy storms and monsoons. Most of this water is lost in floods. A considerable amount of rain also falls in uninhabited areas. The situation is further complicated by the variability in rainfall, since in some years it rains much less than in others. In areas that are naturally dry, these droughts can be disastrous. The droughts that hit Africa during the early 1970s and mid-1980s affected more than 40 per cent of the African population. Many scientists believe that the climate itself is being changed by the build-up of greenhouse gases and that this could increase the frequency of extreme weather with more floods and droughts.

In 1990, the world's reliable run-off provided on average about 1,800 m^3 of water per person. But this is only a global average, which has no meaning to those who live in the arid and semi-arid areas of the world (FAO 1994a).

4.4 Competition for fresh water

During this century, the demand for water has soared with the rapid growth of population, agriculture, urbanization and industrialization. Since 1950, global water use has more than tripled, while per capita use has increased by almost 50 per cent. This has resulted in acute competition among the various users of water as well as among countries sharing the same water resource. The situation is aggravated by the fact that water is still largely a common property resource.

In 1940, total water use was about 1,000 km^3 per year; by 1960 it had doubled, and it doubled again by 1990 when it reached 4,130 km^3 a year. World water use is expected to increase by a further 20 per cent to 5,190 km^3 per year by the year 2000. Agriculture is by far the highest user. In 1990, approximately 70 per cent of the total water usage was for agriculture, while 21 per cent was used for industry (Shiklomanov 1990). The remainder was for municipal water supply and other uses. However, the use by the various sectors varies greatly from country to country. Egypt, for example, uses on average 90 per cent of its annual water consumption for agriculture. On the other hand, about 85 per cent of the water used in Finland is by industry (Figure 4.2).

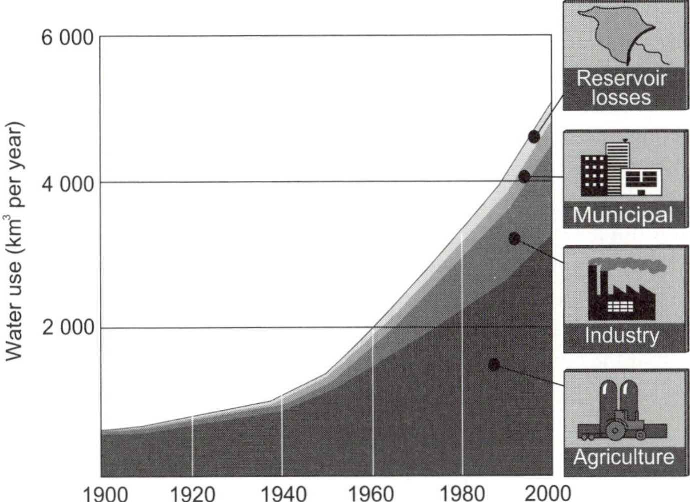

Figure 4.2 Estimated annual world water use by sector.

Many developing countries are now faced with the problem of acute competition between agriculture and municipal uses. Competition over water is serious in a number of cities in China. The problem is particularly severe in and around Beijing and in Tianjin, where water tables have been dropping 1–2 m per year. Pressure is mounting on the farmers to use less water and, considering the inefficiency of water use in agriculture, farmers may be forced to give up a portion of their water. For example, farmers in the Beijing area could lose 30–40 per cent of their current supply within the next decade (Postel 1989). Thus it is imperative that water use efficiency in agriculture should be improved not only to make water available for increasing areas under irrigation, but also to provide water for other uses.

4.5 Water and agriculture

The management of water in agriculture, including fisheries and forestry, has a vital role in the overall use of water at local, national and global levels. Since its inception, FAO has therefore advocated the 'wise use' of water. FAO's prime criterion for agricultural and rural development is 'sustainability'. Thus the approach of FAO to agricultural water use is holistic and embodies the principle of integrated water management.

Accordingly, FAO's policy is not restricted to development of irrigation but embraces all aspects of water management for agriculture and rural development, including meeting the basic water needs of rural people and conserving and protecting the quality of water. About two-thirds of the agricultural harvest comes from rain-fed lands; this is also where the bulk of poverty is found. In an effort to prevent water scarcity from undercutting food production, FAO is exploring a wide range of options such as small-scale irrigation, the reuse of wastewater, water harvesting and improved rain-fed farming within an integrated development approach. The FAO is also concerned with the conservation and restoration of natural resources that underpins sustainable development.

In addition to agriculture, inland fisheries and aquaculture are important sources of food and income in many rural communities, particularly in Africa and Asia. Present levels of production from inland waters, both fresh and brackish, are about 14 million tonnes per year, with over 6 million tonnes coming from capture fisheries and the remainder from aquaculture (FAO 1994a). Inland fisheries will not increase, but substantial gains can be anticipated from aquaculture by the turn of the century. Water quality and quantity are crucial to reliable production and protection of the consumer. Fish production is particularly sensitive to the impact of chemical and organic pollutants which either kill sensitive eggs and young fry or, through the process of eutrophication, rob the waters of essential oxygen. Aquaculture itself is a potential source of water pollution.

The degradation of forest watersheds and the drainage of wetlands have reduced their capacity to retain and slowly release water. The results are accelerated soil erosion, flooding and siltation in dams and waterways. There is an urgent need to restore deforested watersheds, especially in tropical regions, and to bring flooding under control. In the Senegal River valley, for example, the traditional cultivation of the river floodplain following retreat of the seasonal floodwater is being extended by engineer-controlled floods. The result is to increase crop yields while avoiding the environmental damage, high costs and social disruption that full-scale irrigation might cause.

While degradation of the natural environment is having far-reaching effects on water supplies, there are some signs that, combined with industrial and other atmospheric pollutants, it may also be affecting the climate itself. Local changes in rainfall patterns following deforestation have been widely documented, but attention now focuses on the much wider issue of global warming. Over the past 100 years the world has become warmer. This trend is expected to continue, and may even become more pronounced in the future. One result of this global warming would be to make temperate winters wetter and summers drier. The effects in the tropics and sub-tropics are expected to vary, with some areas becoming drier and others wetter.

FAO's Working Group on Climate Change was established in 1988. It has published two position papers on climate change, and is focusing on activities that will clarify the issues and options, or help overcome problems that will be intensified by global warming. Specific studies include:

- the direct effects of increasing levels of greenhouse gases on plant and crop growth;
- how to assess the impact of climate change on irrigated agriculture;
- ways of predicting the regional impact of climate change on crop production;
- mapping and classification of the world's low-lying coastal areas;
- identifying sensitive inshore areas and wetlands;
- monitoring fishery ecosystems;
- studying biofuels as an energy substitute (with a view to reducing the concentration of greenhouse gases in the atmosphere).

4.6 Irrigation and food security

The central role of irrigation in food production and in achieving food security is well-established. Globally, there are now about 235 million ha of irrigated lands, representing about 17 per cent of the total arable area, but production from this small fraction of the land accounts for 35 per cent of the total food harvest. Food production and productivity depend greatly on an assured supply of water. The experience of the green revolution in Asia has demonstrated that the increased use of fertilizer and high yielding rice varieties goes hand in hand with assured water supply. Yields per hectare obtained from irrigated cereals are on average more than twice and often four times as high as on non-irrigated lands.

By far, the largest share of the world's irrigated lands are in the Asia and Pacific region. The distribution on a regional basis of irrigated lands in the developing

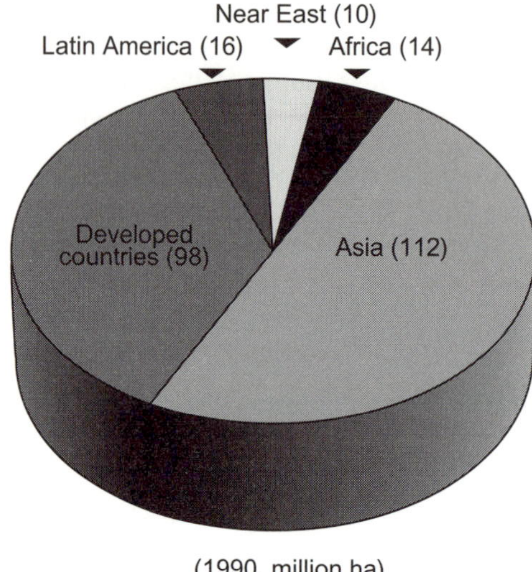

Figure 4.3 A schematic representation of irrigated areas by region.

countries in 1990 was as follows: Asia = 112 million ha; Africa = 14 million ha; Latin America =16 million ha; and Near East = 10 million ha. The total irrigated area in the developed countries was 98 million ha (Figure 4.3).

Much of the irrigation potential on the African continent remains to be tapped. FAO proposes to make a major effort to redress this situation. Droughts have costs beyond the immediate loss of production. They have long-term depressing impacts. Recent studies of drought in the Sahel, the Horn of Africa and southern Africa have revealed the total impact of recurring droughts in terms of famine, loss of lives and damage to the environment. Rational development and improved management of water resources hold the key to achieving food security on the African continent.

4.7 Water resources development and health

Lack of an adequate amount of good quality water for drinking and sanitation is a major cause of human disease, misery and death. According to the World Health Organization (WHO), as many as 4 million children die every year as a result of diarrhoea caused by water-borne infections. None of these deaths would occur if the standards of sanitation that developed countries have long regarded as essential also prevailed in developing countries. Even at the end of the United Nations (UN) International Drinking Water Supply and Sanitation Decade (1981–1990), 31 per cent of the population of developing countries had no access to safe drinking water and 46 per cent were without proper sanitation (Tolba 1992).

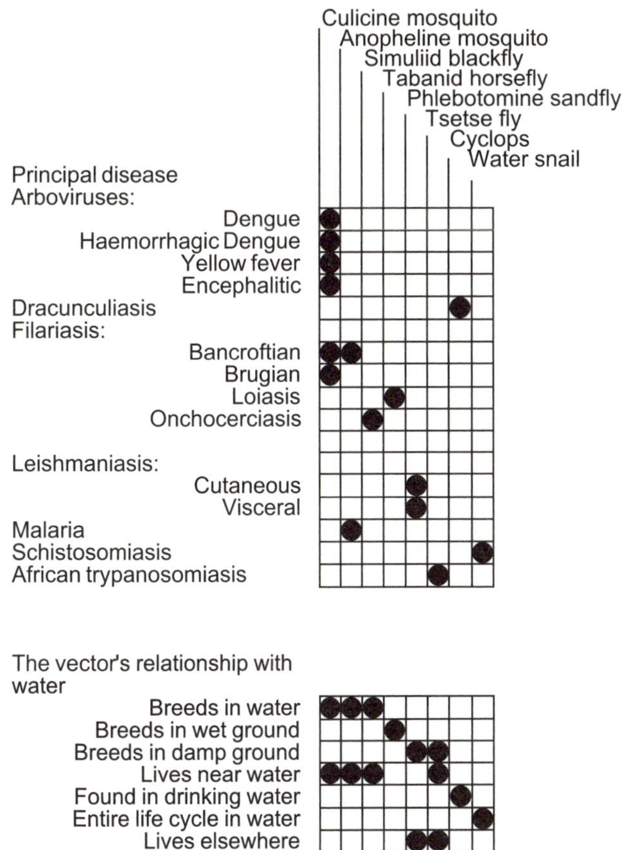

Figure 4.4 Association between vector, disease and water (source: Birley 1989).

In many developing countries, the pollution is so serious that many rivers and drainage canals carry potentially lethal concentration of pollutants and sewage organisms. Traditionally, water itself was used to solve this problem: a polluted stream or river could once be relied upon to cleanse itself every few kilometres. Today, many rivers are polluted from source to estuary: the natural cleansing action just cannot cope.

The problem of water resources development related health hazard is even more serious with regard to vector-borne diseases. The common characteristic of vector-borne diseases is that the pathogen causing the disease is transmitted to man via an insect or snail. These vectors spend part or all of their lives in or near water, but different vectors prefer different types of aquatic habitat as illustrated by Birley (1989) in forecasting the vector-borne disease implications of water resources development (Figure 4.4). Of the many water-related vector-borne diseases, four important ones are: malaria, schistosomiasis, filariasis and onchocerciasis. These diseases are of particular importance because they either:

- cause death and/or severe disability, or
- a large portion of the population at risk becomes ill, or
- they are particularly difficult to control once they become widespread or when they are endemic, or
- the resulting ill health lasts a long time.

Malaria is by far the most important of these diseases, both in terms of the number of people infected annually whose quality of life and working capacity are reduced, and in terms of the death rate from it. Worldwide, some 2 billion people live in areas where they are at risk from malaria and the total number of cases is estimated at 100 million.

Schistosomiasis is a disease that is as widespread as malaria but rarely causes immediate death. An estimated 200 million people are infected and transmission occurs in some 74 countries.

The transmission of dengue viruses are exacerbated by burgeoning urbanization, rapid air travel and the failure of municipal governments to provide adequate and reliable piped water. Mosquitoes, particularly *Aedes aegypti*, breed in artificial containers often made more plentiful by ineffective or non-existent solid waste management programmes. Dengue viruses and dengue haemorrhagic fever have been recorded in 92 countries with two-fifths of the world's population at risk (see Chapter 15).

Improper water resources development and badly designed and managed irrigation and drainage projects have aggravated the water pollution and vector-borne diseases problems in many developing countries. These, combined with the absence of good quality drinking water and sanitation, pose a serious threat to human health and sustainable development.

4.8 Water resources development and ecological consequences

Each year, some 3,300 km^3 of water are removed from the earth's rivers, streams, and underground aquifers to water crops. Practised at such a scale, irrigation has a profound impact on global waterbodies and on the crop land receiving it. Waterlogged and salt-affected lands, declining and contaminated aquifers, shrinking lakes and the destruction of aquatic habitats combine to hang a high environmental price on irrigation. Mounting concern about this damage is making large new water projects increasingly unacceptable.

By far the most serious damage stems from waterlogging and salinization of the soil. It is difficult to put a figure on how much land suffers from salinization worldwide, but estimates from the world's top five irrigators (Table 4.1) clearly

Table 4.1 Area affected by salinization in the top five irrigated countries, 1985–1987 (est.)[a]

Country	*Area damaged (million hectares)*	*Share of irrigated land damage (per cent)*
India	20.0	36
China	7.0	15
USA	5.2	27
Pakistan	3.2	20
Soviet Union	2.5	12
Total	37.9	24
World total	60.2	24

[a] Estimates from Postel (1989).

suggest that the problem is widespread enough to reduce world crop output (Postel 1989).

In California, excessive irrigation and poor drainage have resulted in another ecological disaster. Scientists have attributed the death, deformity and reproductive failure in fish, birds and other wildlife there to agricultural drainage water loaded with toxic chemicals. Signs of irrigation overstepping ecological limits are evident among the world's rivers and lakes. By far the most dramatic is the shrinking of the Aral Sea in Uzbekistan and Kazakstan. The Aral Sea, once the world's fourth largest lake, now stands sixth because so much of the catchment water has been diverted, without precaution, for irrigation. Of the two main rivers that fed the Sea, one provided no flow at all during the period 1974–86 and the other failed five times during 1982–86. By 2000, if no preventive measures are taken, the Aral Sea may be reduced to two-thirds of its present size.

4.9 FAO policy and strategy on food security and water development

Poverty alleviation, nutrition and food security have been major concerns of the FAO since its establishment in October 1945. Over the years, its broad mandate has been further elaborated and given specific directions by its governing body, the FAO Conference, assisted by the FAO Council and a number of international conferences. Thus a comprehensive concept of development guides FAO action, going beyond purely technical aspects of developing and managing natural resources for agriculture, fisheries and forestry. It calls for transformation of rural life and activities in every respect, economic, social, cultural, institutional, technological and environmental, as a means of alleviating poverty and enhancing the well-being of rural populations. Highlights of FAO's policy relating to food security, water development and environmental protection are presented below.

4.9.1 Food security

The FAO has accorded top priority to assisting member nations to achieve food security and sustainable agricultural and rural development. A Special Action Programme has been initiated to address food security with priority given to low-income food-deficit countries (LIFDCs). The strategy of the Special Action Programme would be oriented towards increasing overall food availability, stabilizing yields and generating employment and income in the agricultural sector. Among the LIFDCs, those in Africa will receive particular attention. Water management in general and irrigation in particular have played and will continue to play a significant role in achieving food security in many countries. The success of the 'green revolution' in Asia was largely due to improved irrigation and water control which enabled high yielding varieties to produce near-potential yields. FAO is committed to ushering in a 'second green revolution' to help the LIFDCs achieve food security.

4.9.2 Water resources development

The policies and programmes of water resources development and management were framed to promote integrated management of natural resources. The Natural Resources Programme of FAO aims to promote an integrated and efficient management of land, water and plant nutrients to meet present and future demands on these resources on a sustainable and degradation-free basis. Within this framework, the water resources policies and programmes are designed to assist the member nations to develop and manage their water resources to achieve food security and sustainable agricultural development. The major thrust of this programme includes:

- review and reform of national and international water resources development and irrigation policies and strategies including legislation and institutions;
- assisting formulation of regional and national irrigation master plans including river-basin development plans;
- advice on irrigation development, improvement and modernization including pilot projects to adapt, demonstrate and transfer improved irrigation management practices;
- strengthening national capacities to plan, develop and manage irrigation development through human resources development and institutional strengthening;
- providing guidance on environmental issues relating to irrigation development such as waterlogging and salinity control, water quality protection, prevention of water-borne diseases and undertaking environmental impacts of irrigation and drainage projects.

FAO's 'normative' and 'developmental' activities are mutually supportive and geared to providing the technical assistance member nations need in the most effective manner. The current work programme includes the following activities.

- Under Water Resources Inventories and Evaluation. A worldwide geo-referenced database on water resources and water use for rural development will be initiated and will be related to soils and climate information. At the national level, it should assist member nations in defining policies for the development and management of water resources. At the global level, it should facilitate the study of the evolution of water resources potential, with special emphasis on agricultural use, water scarcity risks and water quality aspects (pollution, vector-borne diseases). This database should also support forecasts, projections, formulation of strategies and eventually be a key input to international agreements related to shared water resources.
- In accordance with the recommendations of the Dublin Conference and Agenda 21 of the UNCED (United Nations Conference on Environment and Development), the scope of the hitherto FAO-supported International Action Programme on Water and Sustainable Agricultural Development will be broadened to cover the total water needs of rural communities. Work will proceed on two mutually complementing fronts: first, by promoting interagency co-operation under the framework of the ACC Intersecretariat Group on Water Resources; and second as a Special Action Programme on Management of Rural Water Resources. Attention will be given to implementing national and subregional programmes that have already been formulated through mobilization of donor support and strengthening of national capacities.
- Irrigation Development, Rehabilitation and Improvement. This will focus on advisory activities for decision makers and donors on the potentials of improved and adapted irrigation technologies for water saving, which were introduced in China, Pakistan and Yemen. During the 1994–95 biennium, reports based on case studies disseminated relevant information to member countries.
- Irrigation Management and Performance. This will cover assistance to countries in their efforts to improve irrigated agriculture. This will include dissemination of the software CIMIS (Computerized Irrigation Management Information System).
- Water Management Technology Transfer. Under this initiative, the series of training manuals for field extension staff will be completed. Training courses are planned in co-operation with national and international institutions. Co-operation with the World Meteorological Organization will continue, leading to the holding of several short national training courses in the use of meteorological data for irrigation management. The inventory of proven operational techniques and methods in the development and management of water will be updated and further disseminated.
- Beyond these, three new elements have been introduced, including transfers from other subprogrammes. Under Health Aspects of Water Development, the control of vector-borne diseases in irrigation schemes will continue to receive attention under the framework of the Joint WHO/FAO/UNEP/

UNCHS (United Nations Centre for Human Settlements) Panel of Experts on Environmental Management (PEEM), through field activities, notably demonstrations, in-service training and strengthening of extension services. A comprehensive set of guidelines on the control of vector-borne diseases in irrigation and drainage projects will be prepared and distributed.

- Work on Water Quality, Drainage and Salinity Control of Agricultural Lands will continue to promote the prevention and control of water pollution, within the context of sustainable agriculture. Four major areas are: water quality assessment and protection; agricultural chemicals and activities related to water pollution; policies and legislation; and prevention of water pollution. Guidelines will be prepared on the prevention and control of water pollution from agricultural operations.
- Treatment and reuse of wastewater in agriculture, forestry and fisheries will be promoted through pilot demonstrations and training. Regional training courses on the reuse of wastewater will be implemented in collaboration with UNEP and WHO.

4.10 International Action Programme on Water and Sustainable Agricultural Development

FAO has been an active partner in the recent activities of the UN to promote sustainable agricultural development. Its contribution to the International Conference on Water and the Environment held in Dublin in January 1992 and the UN Conference on Environment and Development in June 1992 is well recognized. As a consequence of these conferences and in support of sustainable agriculture and rural development, FAO (1990) launched an International Action Programme on Water and Sustainable Agricultural Development (IAP-WASAD). The objective of IAP-WASAD is to assist developing countries to manage their water resources in an integrated manner to meet the present and future needs of agricultural development. In meeting this objective, the IAP-WASAD will assist national governments and regional institutions in setting priorities concerning the use of water and land resources for agricultural development, in updating their current policies and strategies and formulating and implementing programmes to translate their policies and plans into action. IAP-WASAD has identified five priority areas for action:

- water use efficiency
- waterlogging, salinity control and drainage
- water quality management
- small-scale irrigation
- scarce water resources management.

The programme calls for concerted action at all levels – local, national, regional and global – as well as active participation of all members of the community,

Table 4.2 Summary of IAP-WASAD programme in Egypt

Subprogramme	*Projects*	
I Institutional strengthening	1	National Co-ordination Unit (NCU)
	2	Water pricing
II Water and soil management at farm level	3	Village-level agricultural development in old lands (improved irrigation and agricultural practices)
	4	Strengthening agricultural extension and adaptive research (New Lands Demonstration Farm)
	5	Soil survey of the delta
III Environmental monitoring and protection	6	Monitoring of soil and water quality and creation of National Soil and Water Quality Database
	7	Nubaria/Ismailia canal seepage
	8	Improving rural water supply and sanitation
	9	Water-borne diseases
IV Use of marginal quality water	10	Reuse of treated sewage water
	11	Reuse of drainage water
	12	Protection of nitrate pollution of ground water
V Management of scarce water resources	13	Conjunctive use of ground water
	14	Protected agriculture
VI Development of fisheries and ancillary services	15	Marine fish hatcheries
	16	Lake Nasser fish stocking and methods of fishing
	17	Marketing/co-operatives

especially women and international groups such as the UN system, international institutions and bilateral donors.

Since the launching of the programme (FAO 1992), national and subregional programmes on IAP-WASAD have been formulated in Egypt, Indonesia, Mexico, Syria, Tanzania, Turkey, Zimbabwe and in the Lake Chad basin. Out of the seven national WASAD programmes, implementation has begun in four countries: Egypt, Indonesia, Turkey and Zimbabwe. The Egyptian programme is comprehensive and is summarized in Table 4.2.

4.11 Panel of Experts on Environmental Management for Vector Control

The joint WHO, FAO, and UNEP Panel of Experts on Environmental Management for Vector Control (PEEM) was established in 1981 as a result of a memorandum of understanding signed by the heads of the three UN agencies. In 1990, UNCHS joined the programme as a full partner. The objective of PEEM is to strengthen collaboration between the participating organizations and to promote collaboration between them and other appropriate international and national agencies in their programmes and projects relating to the development of natural resources, agriculture, urban water management and health promotion, and in the use of environmental management measures for the control of disease vectors and the protection of human health and the environment.

The PEEM consists of 38 panellists specialized in the fields of water resources, irrigation, agronomy, tropical medicine, entomology, epidemiology, environmental sciences and ecology. It has also established 12 collaborating centres, which are centres of excellence in areas of research that are relevant to the objectives of PEEM. PEEM publications, newsletters, technical meetings and national workshops have contributed to an increased awareness of the vector-borne disease problems associated with water development projects among member nations. In a number of countries, vector-borne disease impact assessments of irrigation and water resources have been evaluated and appropriate measures have been incorporated in the design and operation of irrigation projects.

4.12 Final thoughts

Scarcity and misuse of water pose a serious and growing threat to food security, human health and well-being, industrial development and the ecosystems on which they depend. FAO's policies and programmes recognize the vital role of water in sustaining and supporting life, focusing in particular on sustainable agriculture, rural development and better water management. Such deficiencies can have a dramatic impact not just on agricultural production, but on all aspects of everyday life, very often adding to the burden of women.

The 107th Session of the FAO Council (FAO 1994b), also pointed out that it was necessary to address issues related to physical and economic access to food at the household level as well as at the national level. As a follow-up to the Rio Summit, the Council endorsed the setting up of a Department of Sustainable Development.

FAO will continue to play a leading role in fresh water management. Considering that food production and rural development consume by far the largest portion of available water supplies, effective management of water in the agriculture sector holds the vital key to solving global water problems.

Sustainable development of water resources remains the cornerstone of future development for many countries. The task ahead is indeed immense; but it is not insurmountable provided that appropriate policies, strategies and programmes are adopted and implemented at all levels and by all parties. We should reflect on the realities and opportunities of our time: with more than 800 million hungry people in the world, our greatest challenge is to feed the poor, the hungry and the malnourished of this earth. The ultimate goal is to ensure a twenty-first century where water is not scarce for anyone and food is plentiful for all.

References

Birley, M.H. (1989) *Guidelines for forecasting the vector-borne disease implications of water resources development*. Joint WHO/FAO/UNEP Panel of Experts on Environmental Management for Vector Control, WHO/VBC 89.6, World Health Organization, Geneva.

Clarke, R. (1991) *Water, the International Crisis*. Earthscan Publications, London.

FAO (1990) *An International Action Programme on Water and Sustainable Agricultural Development*. FAO, Rome.

FAO (1992) *Country and Sub-Regional Action Programmes formulated. Undertaken under the International Action Programme on Water and Sustainable Agricultural Development*. FAO, Rome.

FAO (1994b) *Report of the Council of FAO, 107th Session*. FAO, Rome.

FAO (1994a) *Water for Life*, World Food Day issue paper. FAO, Rome.

Postel, S. (1989) *Water for Agriculture: facing the limits*. Worldwatch Paper 93, Worldwatch Institute, Washington, DC.

Shiklomanov, I.A. (1990) Global water resources. *Journal of Water Resources*, **26** (3).

Tolba, M.K. (1992) *Saving our Planet*. Chapman & Hall, London.

Chapter 5

Context and principles of environmental and health impact assessment

Christine E. Ewan, G. Dennis Calvert and Keith W. Bentley

5.1 Introduction

This chapter summarizes the format and subsequent implementation of Environmental Impact Assessment and Health Impact Assessment (EIA/HIA) in Australia. The protocol for environmental and health impact assessment (E&HIA) was designed to be incorporated within existing EIA processes and while the framework is comprehensive, it is intended to be applied selectively. It is neither feasible nor desirable for all E&HIA to follow an inflexible standardized protocol for assessment of health impacts.

The framework is based on the following principles.

- Healthy environments and healthy populations are interdependent. An ecological approach that assesses the impacts of development on human habitats is as necessary for the protection of human health as environmental impact assessment is for the protection of flora and fauna.
- Social justice is a key consideration in public health policy and decision-makers have a responsibility to involve potentially affected communities in decisions that have an impact upon the health and amenity of their environment.
- Decisions should err on the side of caution when impacts on health and the environment are not clearly understood.
- As well as formulating key objectives, indicators and processes for E&HIA, it is also necessary to identify vulnerable groups and their health risks and to ensure that social justice factors that underlie these risks are taken into account in development policy and decisions.

5.2 Steps in conducting environmental and health impact assessment

Each step in the framework is not intended to stand alone, but to be integrated into the relevant part of the existing EIA process.

Step 1: Screening	Does this project need an E&HIA?
Step 2: Scoping	What issues must be addressed in the E&HIA?
Step 3: Profiling	What is the current status of the affected population and the local environment?
Step 4: Risk assessment	What are the risks or benefits? Who will be affected?
Step 5: Risk management	Can risk be avoided or prevented? Are better alternatives available? How can benefits and risks be costed? How can differing perceptions of cost and benefit, nature and magnitude be mediated? Will predictions of future health risk be robust enough to withstand legal and public scrutiny?
Step 6: Implementation and decision-making	Does the assessment provide sufficient, valid and reliable information for decision-making? Is there a conflict to be resolved? How will conditions be enforced? How and by whom will impacts be monitored? How will post-project management be resourced?
Step 7: Monitoring, environmental and health auditing, post-project evaluation	Is the project complying with its conditions? How well are those conditions achieving the desired outcomes? How well is the E&HIA process as a whole achieving its aims of protecting the environment and health?

Multisectoral involvement is essential in planning and assessing each step in the impact analysis. Consultative committees need to be convened in the earliest stages to identify key issues to be addressed by the E&HIA. This facilitates consensual rather than confrontative approaches to decision-making and development.

5.2.1 Step 1: Screening – Does this project need an E&HIA?

Screening aims to identify, as early as possible, those projects that have potentially significant impacts, and that should be subject to E&HIA.

Screening consists of the following steps.

- Establishing contact with community networks.
- Rapid assessment:
  - against a checklist of criteria for public health risk which establish the need for E&HIA;

- in the light of local demographic data, to identify vulnerable populations;
- in the light of local environment data, to identify vulnerable ecosystems where damage has the potential to cause health effects;
- of public interest and/or concern;
- of contribution to disturbances to the global environment.

Not all potential impacts on health are negative. Many developments, such as improved waste disposal or water supply facilities, may result in benefits which outweigh potential adverse effects. It is as important to identify the potential positive health impacts as the negative ones, and this may make a significant contribution to the cost–benefit analysis and final development decision. Health authorities should be responsible for providing a checklist of key potential health effects in each specific proposal for rapid assessment.

An element of judgement will always enter into decisions at the screening phase. The checklist in Table 5.1 identifies key concerns for public health that should be considered.

Table 5.1 Projects that should be screened by health authorities

Developments, activities or policies that:

- generate desirable health impacts such as water supply or waste minimization/disposal
- may cause substantial changes to the demographic or geographic structure of a community
- have the potential to expose workers to hazardous products and processes in the workplace
- open settlements or areas that have the potential to damage ecology, reduce air or water quality to an unacceptable level, introduce disease vectors or parasites, or render recreational facilities and water resources unsafe
- impact significantly on populations that are already vulnerable to ill health (e.g. rural Aboriginal communities, retirement communities)
- impact significantly on land productivity for horticultural and pastoral activities and are potentially detrimental to the socioeconomic status of dependent communities
- affect the microbiological or chemical safety of food chains and food supplies
- substantially increase the demands on public amenities and infrastructure
- involve the storage, transport and disposal of hazardous or toxic materials or wastes, including those that are clinical and infectious
- produce particulate, gaseous or liquid emissions, a significant volume of solid waste, or noxious substances (e.g. radiation)
- generate significant noise, traffic flow, or risk of injury
- generate a high level of community concern (e.g. airports, motorways)

5.2.2 *Step 2: Scoping – What issues must be addressed in the E&HIA?*

Scoping identifies the issues to be considered in depth and those to be eliminated, and it allocates responsibilities for preparing the assessment (Ross 1991). It offers the first opportunity for community participation in the assessment and planning of a proposed development. Scoping teams should be constituted from proponents, groups of experts and representatives of communities or workers who will be affected by the proposal, as it gives stakeholders the opportunity to identify areas on which the E&HIA should focus. Scoping involves the following steps.

- Identify key stakeholders.
- Identify potential health and amenity concerns.
- Determine whether modifications or alternatives to the proposal need to be considered.
- Provide guidelines for the proponent specifying:
  - goals for impact assessment and criteria for monitoring;
  - minimum information requirements;
  - nature and location of relevant existing health data;
  - relevant biotechnical standards and inventories;
  - relevant social and cultural impact indicators;
  - relevant biological and social assessment methodologies;
  - recommended consultation strategies;
  - recommended project implementation, evaluation and monitoring protocols;
  - local agency requirements and policies.
- Determine the responsibility of proponents, health agency and other stakeholders in the preparation and assessment of the E&HIA.
- Negotiate costs and cost-sharing.

5.2.3 *Step 3: Profiling – What is the current status of the affected population and the local environment?*

Profiling is the process of establishing the baseline condition of relevant parameters of the affected community so that likely impacts can be predicted and subsequently monitored. Many of the data for profiling should be available from local health and planning authorities. Not all of the information listed in Table 5.2 will be needed in every case, and much of it may not be available. The list should be viewed as an ideal towards which developing data sets could be targeted. It is not implied that proponents be responsible for compiling those data sets but should refer to them where they are available, relevant and identified by scoping teams.

Table 5.2 Baseline information that contributes to profiling

- Characteristics of the populations of the region, including size, age structure, socioeconomic status, groups at risk
- Physical characteristics of the region, such as frequency of atmospheric inversions, variability of river flow, orientation of prevailing winds, water-tables, geological stability
- Existing land uses, especially those that can be considered incompatible with or inappropriate to the proposed development
- Current health status of the population, including morbidity, mortality, and social health indicators; (data collection using sentinel populations and co-ordinated by a central body would be useful here)
- Current levels of pollutants and environmental degradation
- Existing data and studies concerning the types of problems likely to arise from the development; the reliability of these data needs to be carefully considered, especially if they relate to a region or population that differs from that for which E&HIA is being carried out
- Existing living conditions of the population, especially in relation to factors such as access to water and food supplies, and access to health facilities

5.2.4 Risk analysis

Risk analysis plays an important but somewhat controversial role in environmental health. Currently, it is the primary tool for providing quantitative data aimed at identifying and mitigating environmental health hazards. In any risk analysis, however, there is a number of points at which risk to human health can only be inferred, and both scientific judgements and policy choices may involve selecting from a number of options.

Risk analysis involves both the technical procedures of quantitative risk assessment and the social processes of negotiating criteria for acceptability of proposed policies and developments. To date, most impact assessment has been deficient in its incorporation and assessment of the non-technical components, and in its acknowledgment of the limitations of the technical components.

Methods have yet to be refined for assessing the externalities, human and environmental, as well as the monetary costs, and incorporating them into cost–benefit decisions. Methods for ensuring genuine consultation with communities and their participation in decision-making are also embryonic. There are broad gaps in the data about dose–response relationships between specific exposures and identifiable population–health effects, and the capacity to predict cumulative health impacts on either a spatial or a temporal basis is minimal.

For these reasons it is critical that, where commercial confidentiality is not a genuine factor, risk analyses and follow-up be regarded as 'action research', providing baseline data and refining methodologies for future impact studies.

Risk analysis is best understood in two phases:

- risk assessment
- risk management.

5.2.5 Step 4: Risk assessment – What are the risks or benefits? Who will be affected?

Risk assessment is the identification of hazards and the estimation of the degree and frequency of risk they pose. This is predominantly a quantitative scientific enterprise. The process will vary with the proposal being assessed and with the availability of data to aid in evaluating options. Hard and fast rules cannot be applied, but use of the checklist in Table 5.3 is recommended. Some of the necessary information should be available from public health plans (if they exist), some will be available from published sources.

Dealing with uncertainty in risk assessment

Risk assessment involves many uncertainties, due in part to the deficiency of available data sets, the uncertainty associated with the data that are available, and the fact that although risk assessment attempts to be rigorously scientific, it inevitably involves value judgements. 'Correct' choices are a matter of perspective, and different stakeholders frequently have different perspectives. In some cases dispute is inevitable. The grounds for dispute, however, can be minimized or clarified by attention to certain details in the communication of risk assessment data. Thus (Harvey 1990):

- assumptions and scientific judgements should be stated clearly;
- the nature and magnitude of uncertainties must be delineated;
- risk assessment and risk management should both be addressed;
- the choice of a particular risk assessment methodology over alternate methodologies must be explained in non-technical language;
- risk estimates should be presented to afford comparison of risks;
- the role of risk assessment in decision-making should be specified;
- the agency responsible for decision-making should be clearly identified and publicly accountable.

5.2.6 Step 5: Risk management – Can risk be avoided or prevented? Are better alternatives available? How can benefits and risks be costed? How can differing perceptions of cost and benefit, nature and magnitude be mediated? Will predictions of future health risk be robust enough to withstand legal and public scrutiny?

Risk management is the process by which a regulatory agency and community decide what to do about the results of a risk assessment. Decision-makers and interest groups develop policy, guidelines or management procedures, based on the results of risk assessment combined with feasibility and economic and

Table 5.3 Checklist of questions and criteria for risk analysis

Risk assessment

Hazard identification: Is there sufficient evidence to suggest that a hazard will occur?

- Analysis of properties of the development/agent
- Epidemiological evidence of risk
- Toxicological evidence of risk
- Case reports

Risk estimation: What is the likely extent of risk? Is the risk new or an increment on existing risk?

- Impacts on environment
- Dose–response or dose–effect relationships
- Population exposure and sensitivity
- At-risk groups
- Land use, lifestyles, consumption patterns
- Route, magnitude, frequency and duration of exposure
- Probability and consequences of accidental events

Risk management

Risk characterisation: Is it the best option? Is it worth the risk?

- Public perception of risk
- Risk–benefit analysis, incorporating health effects and effects on the human population as a cost item in personal and economic terms
- Feasibility of alternatives
- Social, political, and cultural implications of options
- Uncertainties and assumptions
- Social justice and equity

Risk mitigation:

Modifying effects:
- Land-use planning
- The proposal
- The environment
- Exposure
- Effects

Mitigating effects (including compensation), formulating or developing:

- emergency procedures for acute exposure
- contingency measures to facilitate detection and timely response to potential problems
- site-monitoring mechanisms to address project closure, risk removal and environmental restoration
- control measures aimed at building public confidence and trust in the approach taken to project management

Incorporating in project approval conditions:

- recommended modifications
- criteria and methodologies for health impact audits

sociopolitical realities. Risk management involves values, social and political considerations, community concerns and economic judgements, and is subject to the constraints of particular statutes. It entails forming and implementing a strategy for accepting or mitigating identified hazards.

5.2.7 Step 6: Implementation and decision-making – Does the assessment provide sufficient, valid and reliable information for decision-making? Is there a conflict to be resolved? How will conditions be enforced? How and by whom will impacts be monitored? How will post-project management be resourced?

Information required by decision-makers

The aim of E&HIA is to provide the decision-maker with the best possible information on the environmental and health effects of the proposed development, and on the possibilities for preventing or mitigating these effects. The information required includes:

- information about all the predicted impacts of carrying out the development (including mitigation measures and cost–benefit trade-offs);
- identification of those impacts that are crucial to the decision;
- illustration of ways in which the decision might be affected by different judgements about the relative importance of impacts.

Compliance with development conditions

E&HIA approvals should specify conditions and criteria for compliance. Human health indicators should be used where they are available and feasible, in addition to more traditional criteria such as point-source emission limits. One of a project's most significant impacts may be the condition and use of its land after the project is complete. Project closure and site clean-up are thus important considerations in any E&HIA.

Compensation can lessen the impact and hardship that may result from implementation of a project. It should be used in addition to, rather than instead of, mitigation measures. Likely compensatable impacts and the appropriate extent and forms of compensation should be identified as part of the scoping exercise.

5.2.8 Step 7: Monitoring, environmental and health auditing, post–project evaluation – Is the project complying with its conditions? How well are those conditions achieving the desired outcomes? How well is the E&HIA process as a whole achieving its aims of protecting the environment and health?

Monitoring and evaluation are integral components of environmental management since they reveal the accuracy of predictions and the appropriateness of decisions based on E&HIA. Techniques such as environmental sampling, surveillance of a marketed product, ongoing epidemiological studies, and evaluation of newly-available health-risk information are employed to monitor outcomes, while new information about risks may lead to earlier decisions being reviewed (Health and Welfare Canada 1989).

Routine monitoring of general health conducted by health authorities can provide a baseline against which to assess anticipated project outcomes. Ongoing health monitoring is the responsibility of health authorities. E&HIA approval conditions should identify any specific responsibilities for monitoring project-specific indicators (e.g. lead levels in blood) which might be the responsibility of proponents.

Environmental and health auditing is the systematic examination of a project, and of process performance in satisfying and achieving environmental goals. In order to achieve this, the state of the existing environment and the health of the affected population needs to be assessed to provide a baseline against which compliance with an environmental impact statement can be checked. This baseline can also be used to assess the accuracy of environmental and health impact predictions, and the effectiveness of mitigation measures used to reduce the impact of a development upon the environment (Reid 1991). All impact assessments should incorporate a follow-up programme of this type, with particular focus on those aspects which the E&HIA has shown to be associated with uncertainty. The follow-up programme should check the validity and accuracy of the project's forecast impacts and the adequacy and effectiveness of the implementation measures. Where necessary, it must also develop measures to mitigate impacts that had not been foreseen. (Ministère de l'Environnement du Québec 1989). Financial responsibility for such follow-up should be determined on a case-by-case basis and should be specified in project approvals.

Health auditing or monitoring of health impacts must be regarded as a research enterprise as well as an attempt to evaluate E&HIA and specific developments. The existing knowledge base and methods in the field are at a very early stage of development, and all available data should be harnessed to build national and international capabilities. The responsibility for health auditing clearly extends beyond the proponents of specific developments. In some cases approval conditions may specify a proponent's specific responsibility, but the background responsibility for health auditing rests with the public health authorities.

The effective implementation of health impact assessment policies and methodologies depends ultimately on our ability to develop adequate indicators and measurement methods for monitoring population health. Health impact auditing should be seen as contributing to broader goals rather than simply evaluating the implementation of specific developments.

The ability to recognize, characterize and reverse adverse environmental health impacts is the most pressing issue facing health professionals into the next century. The biomedical scientific model which underpins most of our existing health care services has a contribution to make in the development of biological monitoring technology and early diagnostic warning signals. Similarly, planners, geographers and ecologists must continue to refine their methods for environmental monitoring, and community advocates and groups must continue to strive for effective participatory planning. All groups need new multidisciplinary skills to address the problems involved in preventing and monitoring the adverse impacts of development on human health.

5.3 National and international status of E&HIA policy

In developing principles for environmental impact assessment through the aegis of the Australian and New Zealand Environment and Conservation Council (ANZECC), Australia has taken up the principles of E&HIA outlined above to only a minor degree. A Memorandum of Understanding between ANZECC and the National Health & Medical Research Council (NHMRC) is principally designed to ensure that there is no duplication in activities and serves as an information exchange mechanism between the health and environmental sectors. Only limited success has been achieved in factoring health considerations into the environmental debate in Australia but the NHMRC, in further developing strategies for the implementation of HIA, is presently examining the implication of risk and economic factors in environmental health management, particularly with regard to public risk perception and the development of health standards, goals and guidelines (AXIS 1996).

Following on from the Intergovernmental Agreement on the Environment of May 1992 (Intergovernmental Agreement on the Environment 1992) and in particular Schedule 3 of the agreement, which focuses specifically on environmental impact assessment, mirror legislation has been implemented by the Commonwealth, states and territories (Western Australia currently has draft legislation before the House). The jurisdictional legislation provides for extensive community involvement in the development of measures together with a formal regulatory impact assessment process taking into consideration economic and best practice principles. A National Environment Protection Council (NEPC) has been established with responsibilities for development of environmental measures taking into consideration health, social and cultural values. The NEPC has identified as a key priority the negotiating of a Memorandum of Understanding between itself and the NHMRC which would represent a further opportunity to pursue the health sector interests for E&HIA with the backing of the NEPC legislation.

At the national level, Tasmania has formally incorporated HIA into the resource management and planning systems of Tasmania. The Environmental Management and Pollution Control Act 1994 (sub-section 74(5)), proclaimed in January 1996, provides that the Director of Public Health may require an EIA to include an assessment of the impact of an activity on public health. The power delegated to the health agency is of equal standing to that of the original EIA ensuring a comprehensive integration with EIA processes in Tasmania. The E&HIA guidelines have been developed in conjunction with the *Environmental*

Impact Assessment Manual (Department of Environment and Land Management 1995).

The Tasmanian legislation has specific lists of activities for which HIA may be required by reference (level 1 activities) and broader provisions in relation to environmental health detriment (level 2 activities) (Tasmanian Department of Community and Health Services 1996). This second level may be activated by local planning authorities for assessment comparable to the Scheduled Activities. Tasmania has published draft HIA guidelines with a view to developing amendments to the public health and planning legislative frameworks to strengthen and refine the integration of HIA into the resource management and planning system of Tasmania. The guidelines are based on those described in this chapter.

The Tasmanian Health Department has proposed, in collaboration with the NHMRC, to develop HIA further within EIA through a series of refinements including:

- protocols for assessment and liaison with other government agencies and the local government sector;
- to develop industry or activity specific guidelines, codes of practice and standards;
- to develop databases to support effective health profiling of communities;
- to develop monitoring processes, community health indicators and health monitoring models.

Community consultations with a particular focus on E&HIA were held in Canberra, Darwin and Sydney during 1996, with a view to obtaining input from state/territory and local government, public health professionals and the community on what role the Commonwealth health portfolio should undertake in future environmental health policy development.

Common themes that emerged from the consultations are listed below.

- The Commonwealth health portfolio should take a lead role, not necessarily in programme or service delivery, but in reinvigorating the principle that protection of human health is a major desirable outcome in environmental/developmental planning. There needs to be developed a sound infrastructure of information (monitoring/surveillance) on the health of the community to guide environmental/developmental decision-making.
- Better environmental health is cost-effective. The potential for savings in the health care budget is considerable, yet it is difficult to realize resources because the outcomes and potential savings are not immediately obvious. Environmental health should be seen as an 'investment', particularly in the context of the large resources that may need to be deployed to manage low-frequency/high-impact environmental health problems (e.g. environmental factor-driven outbreaks of disease).
- The need for effective communication and community engagement emerged

as a major issue. There appeared to be no specific or transparent mechanism for broad intersectoral involvement in setting priorities to ensure ownership and hence protection of public health.

- Consultations recognized that there are many stakeholders with legitimate interests in participating in the policy framework development. There was broad support for improved recognition of the interdependency of roles between policy and service delivery at the state/territory/local government and community levels. There are different legislative responsibilities both within the Commonwealth and between levels of government e.g. NEPC. There needs to be effective co-ordination. Redefinition of the Commonwealth role would facilitate future co-ordination.
- Aboriginal and Torres Strait Islander (ATSI) environmental health issues were acknowledged to be a high priority and deficiencies in present approaches and delivery had been a significant contributor to the failure to consolidate in a sustainable way advances in ATSI environmental health. This was a good illustration of the challenges and importance of intersectoral co-ordination and co-operation.

The E&HIA framework commissioned by the NHMRC that is described in this chapter has also had major impacts both within other national legislatures and within international bodies such as the World Health Organization (WHO).

New Zealand has brought forward health impact assessment guidelines (Ministry for the Environment 1992; 1994). New Zealand has linked its HIA development with the overarching resources management policy enshrined in the New Zealand Resource Management Act (1991), which sets out the rights and responsibilities of individuals, territorial and regional councils, and central government in relation to the sustainable management of natural and physical resources. The act incorporates policy development, plan preparation and administration to allow for the balancing of the broadest range of community interests and values, including consideration in relation to social and cultural values, the need for equity in New Zealand's present and future community, and economic values and health and safety.

The New Zealand approach, similarly to that in Tasmania, incorporates auditing objectives to facilitate the achievement of both environmental and health goals as being integral to ongoing improvements in the process. Finally the New Zealand HIA (Public Health Commission 1995) approach enshrines within the social fabric a framework guidance for all community sectors including public health service providers, consent applicants, consent authorities and the public who have either responsibility for, or an interest in, environmental quality and the protection and promotion of public health.

At the international level, in parallel with the development of the NHMRC guidelines, the authors of this chapter published a methodology for bringing forward consideration of HIA in aid project development programmes and in national development projects for developing countries (Calvert *et al.* 1993). The methodologies espoused have, within the WHO Western Pacific Regional Office (WPRO), been the basis of a number of national workshops on environmental health impact assessment, including, in 1995, pilot training programmes in Laos, China, the Solomon Islands, Papua New Guinea (PNG) and Cambodia. The training modules were designed to accommodate the present status of EIA in these countries, e.g. in China and PNG the workshop's focus was on the HIA.

In the other training workshops, where EIA concepts themselves were at an earlier stage of development, the EIA was presented as a subset, albeit integrally linked with EIA.

The participants at all workshops were drawn from a range of government sectors and co-ordination and co-operation between sectors was a key focus. Following on from the 1995 pilot studies, it was agreed to develop a number of joint collaborative WHO/WPRO country case studies (e.g. Malaysia, Philippines) that can be used in training programmes.

These and other case studies developed under the WPRO programme on E&HIA were presented at a workshop in Singapore that was jointly organized by WPRO and WHO Headquarters with the Institute of Environmental Epidemiology and the Ministry of the Environment in Singapore, and funded by the United Nations Environment Programme (UNEP) and the British Commonwealth Secretariat. As a result of this workshop, a training package on E&HIA was produced and field-tested in selected countries in 1996.

The outcomes from these WPRO activities are being incorporated into globally applicable materials for a training package being promoted by the WHO's Environmental Health Division in Geneva. Similar initiatives with respect to health assessment are detailed in Chapter 7.

References

AXIS Environmental Consultants Pty Ltd (1996) *Risk and Economics in Environmental Health Standards Setting*. Position paper prepared for Department of Human Services and Health, Canberra.

Calvert, G.D., Ewan, C.E. and Bentley, K.W. (1993) *Methods for Environmental and Health Impact Assessment of Developing Projects*. WHO/WPRO/EHC November 1993, Kuala Lumpur, Malaysia.

Department of Environment and Land Management (1995) *Environmental Impact Assessment Manual*. Hobart, Australia.

Environmental Management and Pollution Control Act (1996) Hobart, Australia.

Harvey, P.D. (1990) Educated guesses: health risk assessment in environmental impact statements. *American Journal of Law and Medicine*, **16**, 399–427.

Health and Welfare Canada (1989) *Health Risk Determination: the Challenge of Health Protection*. Ottawa, Canada.

Intergovernmental Agreement on the Environment (1992) Commonwealth Government, Australian Government Printing Service, Canberra.

Ministère de L'Environnement du Québec (1989) *General Guide to the Environmental*

Impact Assessment of Industrial Projects. Direction des Évaluations Environnementales, Québec, Canada.

Ministry for the Environment (1992) *Scoping of Environmental Effects*. Wellington, New Zealand.

Ministry for the Environment (1994) *Resource Consents and Good Practice*. Wellington, New Zealand.

New Zealand Resource Management Act (1991) Reprinted as on 1 March 1994, Wellington, New Zealand.

Public Health Commission (1995) *A Guide to Health Impact Assessment, Guidelines for Public Health Services and resource management agencies and consent applicants*, Wellington, New Zealand.

Reid, A. (1991) *Environmental Auditing: A Summary*. Northern Territory Conservation Commission Internal Report, unpublished.

Ross, H. (1991) *Environmental Impact Assessment: How to Understand and Resolve the Substantive Issues in a Dispute*. Paper presented to IIR Conference on Environmental Dispute Resolution, 14–15 May, Sydney, Australia.

Tasmanian Department of Community and Health Services (1996) *Health Impact Assessment Guidelines Draft*. Hobart, Australia.

Chapter 6

Health opportunity assessment in water resource development

Martin H. Birley

6.1 Introduction

Water resource development projects always have indirect impacts and these can affect the environment, socioeconomic conditions and human health. In this respect they are similar to all other kinds of development and require an environmental impact assessment (EIA). The EIA procedure has been adopted in many countries in South-East Asia, including Australia, and the broad steps are well known. In theory, health is considered a component of EIA but in practice it is rarely given adequate attention. This can be ascribed both to a lack of political will and a lack of suitable procedures. To redress the balance, health impact assessment (HIA) has been advocated. This chapter describes how HIA can be fitted into EIA and outlines the associated methodology that is evolving. It focuses on the less developed countries, where communicable disease is still a major public health concern. It considers the general problem of health and development before focusing on the problem of vector-borne disease and the water development sector.

Impact assessments tend to focus on the negative impacts of projects and the need for safeguards and mitigation. Health opportunity assessment goes further. In addition to identifying the health risks that require management, it identifies opportunities for actively promoting health. For example, in addition to safeguarding communities from water-associated hazards such as vector-borne disease, there may be an opportunity to promote a safe domestic water supply.

6.2 Assessment procedures

In many countries and donor agencies there is now a statutory requirement for EIA. This requirement is still resisted by political pressures, and impact statements are not always adequate and may be ignored or circumvented. The principal administrative steps are for the project proponent to apply to the regulatory agency for planning permission. The application is accompanied by an environmental impact statement (EIS). The EIS is prepared by an independent consultant. The regulatory agency appoints a subcommittee to appraise the project plans, including the EIS, and issues a permission to proceed subject to appropriate safeguards and mitigation measures. Provision is also made to monitor compliance. Human health can be given adequate attention in this procedure by ensuring that:

- the terms of reference for the EIS explicitly address health issues;
- a public health specialist is included in the project appraisal committee;
- safeguards and mitigation measures for protecting human health receive adequate budgetary provision.

6.3 Assessment methodology

There is still no widely agreed methodology for HIA but the broad steps are: identification of health hazards; interpretation of those hazards as changes in health risk associated with the project; and health risk management. Different interpretations of HIA are evolving, with the main division being between urban/industrialized and rural/natural resource developments.

Identification of health hazards is based on a knowledge of the region in which the project is situated and of the kinds of hazards associated with similar projects. For example, it is well known that vector-borne disease hazards are associated with water resource development projects that are situated anywhere within the tropical and subtropical regions. Some of the other health hazards are less well known. These include sexually transmitted diseases associated with the construction phase; poisoning associated with misuse of agricultural chemicals; injuries associated with increased traffic, fast moving machinery and tunnelling; malnutrition associated with loss of subsistence crops; and mental disorders associated with poorly managed resettlement. So there are five major categories of health hazards to consider.

It is the responsibility of the person undertaking the impact assessment to ensure that all significant health hazards have been identified. This should involve a scoping process in which other agencies and the general public are consulted.

Assessing whether the associated health risks will change as a result of the project is the most important and most difficult component of the analysis. Quantification is usually not practical and a ranking procedure may have to suffice. The ranks adopted could be: no change; reduction in health risk; increase in health risk. Although this is very crude, it may be sufficient to ensure that health risk management is addressed.

In order to estimate this change of health risk, it is necessary to subdivide the problem further. There are many ways of doing so but the following steps have proved of practical value.

- Identify the vulnerable communities.
- Establish the environmental factors that expose the communities to the hazards.

- Assess the capability of health protection agencies to safeguard the communities.

Communities are vulnerable because of their immune status or behaviour. Their behaviour is determined by their economic status, knowledge and attitude towards the hazard. For example, in many rural communities in poor countries people do not know that malaria is transmitted by the bite of certain mosquitoes.

The term environmental factors is also known as environmental receptivity, especially when considering vector-borne diseases. In the case of water resource developments, the environmental factors include the amount and type of surface waters. These may provide the breeding sites for disease vectors. For example, the vector of the deadly disease Japanese encephalitis (JE) likes to breed in flooded rice fields. But the risk of a JE epidemic may depend on the presence of another environmental factor: herds of pigs, which act as the amplifying host. Of equal importance is a further environmental factor: the siting of human communities in relation to the vector breeding sites; most mosquitoes do not usually fly more than about 2 km.

Finally, there are several agencies in addition to the Ministry of Health that are responsible for protecting human health. These include agencies responsible for water supply, sanitation, roads, pesticides, choice of crops, education, compensation and agricultural development. The most important agency is often the Ministry of Health. In rural areas of poor countries they are usually represented by a small health centre. Such a centre is usually understaffed, underequipped and short of medical supplies. It is unlikely to be able to cope with a population influx associated with a development project or the new types of health hazard that may arise.

6.4 Sources of information

Concern about the health effects of water resource development has led four UN agencies to support a joint activity known as the Joint WHO/FAO/UNEP/UNCHS Panel of Experts on Environmental Management for Vector Control (PEEM). This panel has been responsible for preparing and printing a number of guides to assist managers of water resource developments. These include *Guidelines for Forecasting the Vector-Borne Disease Implications of Water Resource Development* (Birley 1991). This document has been released in both paper and electronic formats. It may be downloaded directly from the Internet and is located at http://www.liv.ac.uk/~mhb/index.htm.

Water resource development is but one example of development activities that modify the environment. Policies and procedures are required for all such developments. The driving force for such assessment is often the donor agency, and the Asian Development Bank provides a fine example. The Office of the Environment at the Bank has published a series of guidelines on impact assessment. One of the most recent is *Guidelines for the Health Impact*

Assessment of Development Projects (ADB 1992). Although modelled on the internal decision-making process within the Bank, the procedures described are of more general applicability. Consequently, the procedures have been incorporated in a book entitled *The Health Impact Assessment of Development Projects* (Birley 1995). This contains a wide-ranging review of the health hazards associated with development projects. Many of these health hazards are of interest in the specific case of water resource development. The book will be released in both paper and electronic formats.

There are many other relevant publications that should be consulted by readers who wish to obtain more detailed information. These range from books specifically concerned with parasitic disease (Hunter *et al.* 1993; Oomen *et al.* 1988; Service 1989) to books concerned more broadly with impact assessment (Vanclay and Bronstein 1995).

6.5 Case studies

Health impact assessments and evaluations are usually conducted on a contractual basis and this imposes a right of confidentiality on the data uncovered, limiting the advancement of the subject. In the following examples, care has been taken to hide the precise locations of the project where this has been considered necessary.

6.5.1 Example from West Africa

Benin is classed as a country in greatest need with a low gross national product (GNP) per capita and a population largely engaged in subsistence agriculture. A number of small water resource developments have been built in collaboration with the Food and Agriculture Organization (FAO) of the UN and more are planned. FAO was concerned about the possible health impacts of their projects and requested a rapid health opportunity evaluation of the completed projects. The evaluation was undertaken by a national team with support from European consultants (Akowanou *et al.* 1993; Birley 1993; WHO 1994). Two sites with contrasting health opportunities will be described.

Site A consisted of a women's market garden project that produced vegetables for sale in the capital city. Water was pumped from a nearby permanent swamp into small concrete tanks. The women then filled their watering cans from the tanks and watered the vegetables. The health hazards identified were malaria, schistosomiasis and diarrhoeal diseases. The vulnerability of the community was partly due to their poverty and partly due to an almost complete ignorance of the scientific explanation of the causes of communicable disease. The health risk analysis concluded that there was no additional risk of malaria associated with the project because the system of watering did not create surface pools. At first sight, the concrete tanks suggested an increased risk of schistosomiasis because they would have made ideal bathing sites for children. However, it was characteristic of the local culture to regard such behaviour as inappropriate. The project did not promote the occupational health risk of schistosomiasis through water contact by the cultivators. Indeed, the risk was reduced as irrigation water was not obtained directly from the swamp.

Domestic water supply for the village came from two sources. The first source was the swamp itself: the water was used for bathing and washing clothes with ample opportunity for water contact. The second source of water was an open tube well. In order to draw drinking water for their households, villagers brought their own containers and ropes to the well from the house. So the well water could become contaminated. In addition, the village did not have latrines; the villagers followed the local custom of open defecation in the surrounding bush.

We concluded that although the project did not increase health risks, there were several opportunities to improve health through the medium of the project. The villagers could use the profits obtained from their market gardening to purchase impregnated bednets and reduce the risk of malaria. The village could be provided with latrines as a demonstration project. The agricultural extension officer associated with the market garden could encourage the villagers to consider methods of improving their domestic water supply. The agricultural success of the project could be used to support a community-based health education project using village volunteers.

Site B was in the drought-prone remote north of the country and some 20 km from the nearest health centre. A small rain water dam had been constructed as part of a livestock watering project. During the long dry season, this dam was the preferred source of domestic water, although it was not intended for this purpose. The other sources were two deep wells that tended to dry up and were insufficient for the size of the population.

The health hazards identified were the same as before. During the dry season, the multiplicity of footprints in the soft mud around the reservoir provided the only local breeding sites for malaria mosquitoes. The stones of the dam wall were habitats for the snail host of schistosomiasis. The water itself was heavily contaminated by animal faeces. Domestic water was collected by walking into the reservoir to fill containers. Latrines were absent, open defecation was the norm, the community was illiterate and uninformed about the causes of communicable diseases.

It was concluded that the reservoir increased the risk of malaria by extending the breeding season. It increased the risk of schistosomiasis by providing an ideal breeding site for snails and for water contact. It also increased the risk of diarrhoeal disease by providing a contaminated source of domestic water. Various actions were proposed to manage the health risks and these were the same as the actions required to promote improved health more generally.

Figure 6.1 A cattle watering project in West Africa. During the dry season the reservoir was used by both cattle and people. It was heavily contaminated by cattle faeces. Snail hosts of schistosomiasis were abundant on the dam wall. Trampling of the margins provided breeding sites for malaria mosquitoes. (Source: TROPIX /M. and V. Birley.)

6.5.2 Example from South-East Asia

The second example is also an evaluation rather than an assessment. It was undertaken after a project was operational and determined what actions had followed from recommendations in a feasibility study.

Large dams have been constructed in many parts of South-East Asia to provide the electricity needed for the growing industrial base. They are usually located in remote, forested regions, can flood large areas of land and displace a forest-dwelling subsistence community. The political imperative to resettle displaced communities is now reasonably well established and usually forms part of the terms of reference of the project.

Careful studies had been commissioned to ensure the success of the re-settlement plan. The community to be displaced had a staple diet of upland rice. It was noted that rice was central to their culture and far more significant to them than merely something they ate. The community had been interviewed and asked their preferences for their future homes and livelihood. They had stated that they wished to continue growing their upland rice but would welcome assistance with the development of additional cash crops. This view was recorded and clearly presented in the feasibility report. I visited the community to undertake a rapid health impact evaluation, somewhat informally, a few years after the resettlement was completed. I found that the preferences of the community had been ignored: they had been provided with tree plantations for the production of cash crops and very little land had been set aside for subsistence rice production. The land on which their new plantations were located was of inferior quality to their old farms. It was located on the sides of the hills in place of the now-flooded valley bottoms that had alluvial soils. The community was dissatisfied and could not to afford to buy food. They had tried to re-establish their traditional way of life by cultivating rice among the tree crops. More importantly, they had established new rice farms and temporary shelters deep within the forest reserve.

In South-East Asia, malaria is typically associated with the forest or forest fringe. In this particular community, there had been good malaria control through residual house spraying. This was greatly assisted by the habits of the local malaria vector and the extremely compact nature of the human settlements. The local malaria control unit kept good records of where it sprayed and the annual malaria case rates. Some recent increase in prevalence rate had been noted, but this had been ascribed to illegal incursions of infected individuals from the neighbouring country. Careful examination of the records suggested that the malaria rate had started to increase at the time of dam construction. It was concluded that the agricultural policy adopted for the resettlement scheme was at least partly responsible.

6.6 Economic analysis

There are still many limitations to the methodology of health opportunity assessment. One such limitation is our ability to attach economic costs and benefits to health risks, safeguards, mitigations and health promotion measures. In order to define this problem more closely, two reviews of the economic analysis of health impacts have been commissioned (Goodman 1994; Thomas 1994a; Thomas 1994b).

When health risks increase as a result of development, hidden costs are transferred to the health sector and to the community. The health sector must pay, in part, for health care and treatment. The community must pay, through ill health, for loss of productivity, labour substitution, loss of education and the cost of seeking health care.

The health costs and benefits of irrigated agriculture can be divided into direct and indirect categories. The direct costs include those of the health care provider and those borne by patient and family during periods of illness. Indirect costs apply to loss of earnings and loss of other household activities. Benefits refer to the reduction in expenditure resulting from the project.

The non-market indirect costs of disease are particularly interesting: ‘as long as income losses from endemic diseases are minimized through family labour substitution, the cost of ill health may appear low, and disease eradication may not appear essential to economic growth’ (Nur 1993). In a study of 250 tenant farmers in the Gezira, Nur found that over 8,000 working hours per year were totally lost to malaria. Agricultural output was maintained by family labour substitution but the cost of activities foregone was not valued. Studies of the effect of malaria in Africa have indicated that where the average daily earnings (at 1985 prices) were US$0.20, the direct cost was US$1.80 and the indirect cost was US$8.00 per episode of malaria (Evans and Jamieson 1994).

On irrigated rice projects in Cameroon, malaria was not found to be a significant explanatory variable influencing rice output, but schistosomiasis was significant (Audibert 1986). The market value of indirect costs has been studied in relation to the agricultural wage labour lost to schistosomiasis. The results have been inconclusive, with some studies demonstrating impaired labour productivity while others did not. The general problem of valuing indirect costs is that people are often infected with several different parasites simultaneously. Their loss of productivity cannot then be ascribed to the single infection under study. There are some studies of the direct cost of schistosomiasis control on estates. These studies were used to justify schistosomiasis control, although the conclusions are questionable based on the evidence presented (Goodman 1994).

The International Rice Research Institute has investigated the health cost of pesticide usage in rice production (Pingali *et al.* 1990). They concluded that a 1 per cent increase in pesticide use implied a 0.25 per cent increase in health costs. The health costs include the direct costs of treatment and the indirect costs of lost labour during recuperation from the effects of poisoning.

Several studies have sought to identify the effect of a transition from subsistence to cash-cropping on the nutritional status and health of households (Goodman 1994). The evidence is inconclusive. Household expenditure on health care, education and housing did increase with income, but the effect was small.

There is a number of studies of the cost-effectiveness of vector control. These include comparative costs of chemical and biological control and environmental measures that can largely be charged to the agricultural sector (Bos and Mills 1987; Mitchell 1987). Many of these studies are weak on the analysis of effectiveness. In addition, sensitivity analysis was needed to determine how the conclusion was affected by changes in key parameters such as discount rate (Mills and Bradley 1987).

6.7 Promotion of health opportunity assessment

Despite the relatively large amount of information available, health opportunity assessment is not a routine component of development planning. There is a need for advocacy, training and further case studies. Advocacy is provided by holding workshops and seminars and making presentations at international meetings. These serve to sensitize decision-makers to the problems and procedures. A range of training courses is also required. These should include training for the specialists with public health backgrounds who may be expected to undertake impact assessments and to prepare reports. Good case studies provide them with a model to follow. But training is also needed for the middle-level government officers who are responsible for appraising project proposals and making plans for mitigation measures and monitoring activities. Recent experience of conducting such training courses in Africa is described in the next chapter in this volume. Government officers can do much to promote health opportunity assessment with one key action. This action is the preparation of good terms of reference.

6.8 Terms of reference

The terms of reference for an environmental impact assessment of a project usually contain a phrase like 'and the effect on human health shall be considered'. As a result, there is usually a paragraph or page, somewhere within a completed report of many hundreds of pages, that contains some vague generalizations about health. These will typically refer to water-borne diseases and reiterate the need for providing medicines and health education. If the study has been assisted by a vector biologist, it will usually contain a list of the biting insect species encountered during a baseline survey, with notes on their relative abundance. Such reports do little to promote human health.

The minimum requirement for a health component in the terms of reference of an environmental impact assessment is as follows.

The study should identify the principal health hazards associated with the project and interpret the change in health risk that is likely to occur during the construction and operational phases. In order to analyse the health risk, reference should be made to community vulnerability, environmental factors and capabilities of health protection agencies. A series of recommendations should then be prepared in order to ensure that no significant health risk is increased by the project and that significant pre-existing health risks are reduced. Recommendations should be ranked by cost, acceptability and sustainability.

The principal conclusion of this chapter is that health will continue to be adversely affected by water resource developments in the tropics until terms of reference such as the above are routinely applied, recommendations are produced and action is taken to promote human health.

References

ADB (1992) *Guidelines for the Health Impact Assessment of Development Projects*. Environmental Paper No. 11, Asian Development Bank, Manila.

Akowanou, E., Akogbeto, M. and Goussanou, B. (1993) *Évaluation de l'impact des amenagements hydro-agricoles et pastoraux sur la santé des populations au Bénin*. FAO BEN/91/002 et BEN/88/012, Rapport principal, FAO, Rome.

Audibert, M. (1986) Agricultural non-wage production and health status: a case study in a tropical environment. *Journal of Development Economics*, **24**, 275–291.

Birley, M.H. (1991) *Guidelines for Forecasting the Vector-borne Disease Implications of Water Resource Development*. Joint WHO/FAO/UNEP Panel of Experts on Environmental Management for Vector Control. WHO/CWS/91.3, second edition, WHO, Geneva.

Birley, M.H. (1993) *Benin Mission Reports*. Liverpool School of Tropical Medicine, HIP/93.08 and HIP/93.15.

Birley, M.H. (1995) *The Health Impact Assessment of Development Projects*. HMSO, Norwich.

Bos, R. and Mills, A.J. (1987) Financial and economic aspects of environmental management for vector control. *Parasitology Today*, **3**, 160–163.

Evans, D.B. and Jamieson, T. (1994) Economics and the argument for parasitic disease control. *Science*, **264**, 1866–1867.

Goodman, H. (1994) *Economic analysis of the health impact of irrigated agriculture, a review*. Liverpool School of Tropical Medicine, Health Impact Programme. Commissioned report, HIP/94.27.

Hunter, J.M., Rey, L., Chu, K.Y., Adekolu-John, E.O. and Mott, K.E. (1993) *Parasitic diseases in water resources development*. WHO, Geneva.

Mills, A.J. and Bradley, D.J. (1987) Methods to assess and evaluate cost-effectiveness in vector control programmes, in selected working papers prepared for the 3rd, 4th, 5th and 6th meeting of the Joint WHO/FAO/UNEP Panel of Experts on Environmental Management for Vector Control. WHO/VBC 87.3, WHO, Geneva, pp. 134–141.

Mitchell, C.J. (1987) The cost-effectiveness of environmental management as a vector control measure, in selected working papers prepared for the 3rd, 4th, 5th and 6th meeting of the Joint WHO/FAO/UNEP Panel of Experts on Environmental Management for Vector Control. WHO/VBC 87.3, WHO, Geneva, pp. 126–133.

Nur, E.T.M. (1993) The impact of malaria on labour use and efficiency in the Sudan. *Social Science and Medicine*, **37**, 1115–1119.

Oomen, J.M.V., de Wolf, J. and Jobin, W.R. (1988) *Health and Irrigation: incorporation of disease-control measures in irrigation, a multi-facted task in design, construction, operation*. International Institute for Land Reclamation and Improvement, Wageningen.

Pingali, P.L., Marquez, C. and Palis, F.G. (1990) *Health costs of long-term pesticide exposure in the Philippines – a medical and economic analysis*. Social Science Division Papers, International Rice Research Institute, Los Banos, Philippines, pp. 90–104.

Service, M.W. (1989) *Demography and Vector-Borne Diseases*. CRC Press Inc., Boca Raton, FLA.

Thomas, M. (1994a) Economic analysis of the health impact of development projects: a bibliography. Liverpool School of Tropical Medicine, Health Impact Programme. Commissioned report, HIP/94.04.

Thomas, M. (1994b) Economic analysis of the health impact of development projects: the theoretical basis. Liverpool School of Tropical Medicine, Health Impact Programme. Commissioned report, HIP/94.07.

Vanclay, F. and Bronstein, D. (1995) *Environmental and Social Impact Assessment*. John Wiley & Sons, New York.

WHO (1994) *Développement Agricole et Santé au Bénin: Rapport d'un séminaire national*. Joint WHO/FAO/UNEP/UNCHS Panel of Experts on Environmental Management for Vector Control, WHO/CWS/94.1, WHO, Geneva.

Chapter 7

Health opportunities in water resources development: a course promoting intersectoral collaboration

Peter Furu, Martin H. Birley, Charles E. Engel and Robert Bos

7.1 Introduction

Development of natural resources always results in environmental change, it frequently causes demographic change and sometimes leads to environmental degradation. These immediate impacts may imply indirect, adverse effects on the health status of the affected communities. In the past, the health ministry was expected to pick up the pieces: development projects therefore often led to an increased burden on the health sector. Recently, it has become accepted that these health implications should be a matter of concern to a range of different government sectors. In institutional terms, these include ministries responsible for the natural resources and their development, ministries of planning and finance, ministries of health and environment, national environmental councils, ministries of local government and the counterparts of all these ministries at lower administrative levels, such as provinces, regions and districts.

Ministry staff responsible for policy-making, development planning and programme implementation are usually specialists who are unfamiliar with the concerns of colleagues in other sectors. Communication channels between staff of different ministries, be they formal or informal, are usually limited and not very effective, and intersectoral competition often prevents a rational sharing of resources even if development touches the interests of various sectors.

Intersectoral collaboration has long been promoted by the health sector as an essential element in achieving its goals, globally embraced in the World Health Organization's (WHO) 'Health For All' strategy. The need for institutional restructuring and capacity building in support of intersectoral collaboration in the environment and development context was a key issue in the Brundtland report (WCED 1987). Agenda 21, the blueprint for sustainable development in the twenty-first century adopted at the Rio de Janeiro Earth Summit (UN 1993), highlights the promotion of education and training in the same context and explicitly refers to the need for innovative training methods. In the more limited context of water resources development associated vector-borne disease problems, the joint WHO/FAO/UNEP Panel of Experts on Environmental Management for Vector Control (PEEM) recommended, as a result of technical discussions at its eighth meeting (education and training for the planning, design and implementation of environmental management for vector control), various training approaches together with appropriate follow-up (WHO 1988). These recommendations on training and the current status of the follow-up are as follows.

- 'WHO, FAO and UNEP should encourage governments and financing agencies to increase intersectoral and interprofessional collaboration in the control of vector-borne diseases by establishing specific policies in relation to the education of personnel involved in the planning, implementation and operation of water resources development projects.'

  All three agencies have made increased efforts to include health in national development policies and strategies. Particularly noteworthy are the efforts of FAO to include environment and health issues in national water policies and national strategies for integrated water resources management, as developed under its Action Programme for Water and Sustainable Agricultural Development; the effects of WHO to ensure that health is effectively included in national plans for sustainable development; and the efforts of the Liverpool School of Tropical Medicine Health Impact Programme (a PEEM Collaborating Centre), which has organized a number of seminars in the eastern Mediterranean on the health impact of development policies.
- 'WHO, FAO and UNEP should collaborate in the development of strategies to incorporate environmental management for vector control into agricultural extension programmes.'

  Under the PEEM umbrella three inter-regional workshops have been held (Alexandria in 1991, Bangkok 1991 and Tegucigalpa, Honduras 1992) on the promotion of environmental management for disease vector control through agricultural extension programmes. The second phase of this initiative foresees the production of guidelines for extension workers and the implementation of country action plans that resulted from the workshops.
- 'The panel should collaborate with appropriate agencies in commissioning general guides for those who are asked to organize or implement educational sessions or entire programmes relating to intersectoral collaboration in vector-borne disease aspects of water resources development. Each guide would concentrate on education or training of a specific audience. . . . Educational principles and their suggested application would be presented with illustrative examples from real life experiences.'

  With a focus on engineering schools, and recognizing the fact that 'special pleading' may meet with considerable resistance in those establishments where curricula are already overcrowded, the panel decided to proceed with the development of a modular self-learning package on environmental engineering and management for disease vector control for engineering students. This activity awaits external funding.
- 'The panel should commission and field test a number of educational

packages designed to foster attitudes and general competence for intersectoral collaboration in institutes . . . where the spectrum of interdisciplinary expertise may be limited.'

In further discussions this idea was modified to become the proposal for the training course on intersectoral collaboration for middle level staff of relevant ministries that is discussed in this chapter.

- 'The panel should . . . organize, on a regional basis, short, international workshops for high level decision-makers including financial planners.'

  Since 1988 PEEM has organized three national seminars on water resources development and vector-borne diseases in Kenya (1988), Benin (1993) and Zambia (1995), which have contributed significantly to a greater awareness among decision-makers in various relevant sectors.

7.2 Considerations leading up to course development

At the eleventh PEEM meeting in 1991, the Danish Bilharziasis Laboratory (a PEEM Collaborating Centre) presented a proposal for a comprehensive training course on intersectoral planning of water resources development. The proposal was approved by the panel and the activity was included in its programme of work, with the aim of strengthening the capacity of national officials in relevant ministries to include in their assignments the potential impact of development projects on the health status of the construction labour force, of the affected communities and of formal and informal migrants that may be attracted by the economic activity such projects cause.

Development planners are accustomed to working in the framework of the project cycle (Tiffen 1989). During the phases of this cycle there are several opportunities to ensure that human health is not only safeguarded, but actively promoted as well. Crucial decision-making moments include the formulation of terms of reference for the project's feasibility study, the technical and economic appraisal of environmental and health impact studies, and the negotiations over the financing of the project and the allocation of funds to specific subcomponents, including mitigating measures for environment and health. Some of the basic principles in this connection include the following.

- A health impact assessment must be part of the environmental impact assessment. It should not be carried out in isolation.
- Human health should be considered in a cross-cutting way, in the context of an environmental impact assessment, and not, as is often the case, in a sectoral way with recommendations addressing exclusively the health authorities.
- Rather than an **impact** assessment, with a focus on adverse health effects, it should be an **opportunities** assessment, which identifies ways to improve the health status as part of the development process.

- The first line of interventions should be environmental engineering and management measures, as an integral part of the project, to reduce health risks to a minimum and to keep the subsequent burden on the health services within manageable proportions.

Recognition of the need for training in the skills of intersectoral collaboration and negotiation led to an agreement in principle between PEEM, the Danish Bilharziasis Laboratory (DBL) and the Health Impact Programme (HIP) of the Liverpool School of Tropical Medicine to embark on the design of the course structure and contents, on the development of course materials and to run a number of trial courses. The course would adhere to the principles listed above and relate to the project cycle concept.

Table 7.1 Health Opportunities in Water Resources Development – features of the courses

Date and venue	*Duration (days)*	*Number of participants Male*	*Female*	*Number of ministries represented*	*Outputs*	*Follow-up*
Darwendale, Zimbabwe 7–18 September 1992	12	27	1	5 + university + municipal authorities	Rapid assessment of the Mupfure Irrigation Scheme	Yes: in-depth impact assessment; Mupfure policy study; follow-up seminar
Akosombo, Ghana 26 January to 11 February 1994	17	22	2	7 + Env. Prot. Council + Volta River Authority	Generic TORs, planning flowchart	Yes: follow-up seminar
Arusha, Tanzania 13–30 March 1995	18	12	7	6	Generic TORS, planning flowchart	No
El Zamorano, Honduras 9–28 June 1996	18	17	7	5 countries; 4–5 from each	Generic TORs, planning flowchart	No
Aurangabad, India 10–28 November 1997	18	19	1	4 states; 3–6 from each	Generic TORs, planning flowchart for 4 states	Yes: forward TORs to each state; possible institutionalization of course

From the start, the organizers visualized clearly the end product: a set of training materials for a course aimed at an audience of mid-level managers of ministries involved in water resources development planning, health and environment. The course did not aim at turning the participants into mini-experts in each of the fields related to the subject, but was directed at developing their skills in intersectoral decision-making. The materials allowed course implementation by national institutes without major external inputs. It was further agreed that the course would be tested five times before completing the materials. For each course, clear development objectives would be defined (in addition to the specific capacity building objectives intrinsic to the course). The international organizers closely monitored and evaluated each course through preparatory and evaluation meetings before and after, and through evaluation exercises during the courses. An overview of the main features of each of the courses is given in Table 7.1.

The remainder of this chapter will describe the course development in detail, together with the evolution in aims and learning objectives. The course reports are available from the PEEM Secretariat at WHO, Geneva (WHO 1993; 1994; 1995; 1996; 1998).

7.3 Course development and learning objectives

The first course in Zimbabwe followed conventional lines. During the first week information transfer through lectures by overseas experts covered five themes:

Day 1: General introduction, policy issues and planning procedures
Day 2: Forecasting monitoring and surveillance
Day 3: Environmental engineering problems
Day 4: Economic and financial aspects: cost-effectiveness analysis
Day 5: Strengthening of health services.

For each of these themes learning objectives were formulated and the overseas experts were asked to develop modules accordingly. Only on Day Two did the programme divert from the routine lecturing approach: participants were divided into small groups and were involved in an active learning approach on the subject of health impact assessment. The second week concentrated on the application of the newly acquired knowledge to a real project in the planning stage: the Mupfure Irrigation Scheme. At the end of the week, groups had to present the outcome of the exercise, and in addition they were subject to a Modified Essay Question test.

The evaluation of this first course showed that it had stayed too close to conventional approaches of passive knowledge transfer and had therefore not met its objective of developing skills rather than making each participant into a mini-expert in all relevant fields. The participants themselves indicated that the first week's programme had been overloaded, except for the problem-based learning day, about which they were enthusiastic. They also expressed their satisfaction with the group work on a real life project in the second week, but felt that this should have been more structured. From the efficiency point of view it was clear that a heavy reliance on overseas experts would not meet the criteria of replicability and sustainability once countries implemented the course themselves.

Based on this experience, particularly the positive feedback on Day Two of the Darwendale course, the organizers decided on a complete overhaul of course contents and structure, adopting a problem-based learning approach (Engel 1992) for its entire duration. In this revised set-up the aim of the course was, more than before, to foster participants' confidence in colleagues with different disciplinary backgrounds, working in different ministries, to demonstrate that intersectoral collaboration is a feasible option, to develop negotiating skills and to demonstrate that the intersectoral process is efficient and leads to desirable outputs.

The international organizers defined the crucial decision-making moments in the project cycle and developed tasks for which learning objectives were formulated. For each task, the process was broken down into small steps, permitting the participants to discuss options and priorities in small, intersectoral groups and reach a consensus. In the *Task Guides*, which were developed first for the 1994 course in Akosombo, each step was covered on a page and groups worked through each task page by page. A more detailed description of these guides is presented below. The task-oriented, problem-based learning approach was reported by Birley *et al.* (1995; 1996). Initially, four tasks were identified as follows.

- An initial health examination at the project identification/pre-feasibility stage requiring a rapid impact assessment. Output: a recommendation of whether or not an in-depth health impact assessment is necessary as part of the environmental impact assessment.
- Appraisal of the health impact assessment report. This includes the technical appraisal of a complete health impact report and the economic evaluation of the intervention options it proposes. The appraisal provides the basis for negotiations with external donors and for the formulation of an action plan for intersectoral implementation, monitoring and surveillance. Outputs: an appraisal report presenting plausible and justified options for the implementation of health protective and promotional measures to be included in the development project.
- Formulation of generic terms of reference based on the experience gained from the technical and economic appraisal carried out in Task Two. The focus should be on the translation of hazards related to community vulnerability and environmental factors into perceived risks that take into account the capacity of the health services to deal with them. The terms of reference (TORs) should also refer to health opportunities in addition to health risks. Output: generic terms of reference.

- Formulation of an intersectoral action plan that outlines the institutional arrangements, implementation schedule for planned activities and monitoring/ surveillance during the construction and early operational phases. Outputs: a plan containing a time schedule, a memorandum of understanding indicating sectoral responsibilities and resource sharing, and a monitoring/surveillance component.

This approach and the materials developed required a different course organization. To start with, the course was extended to 17 days (the third and fourth courses to 18 days), and several features were added to facilitate the task-oriented group work including the provision of non-expert tutors, the use of local experts, the establishment of a provisional library and the organization of field trips. Similarly, the course programme was adapted to the new learning approach. Routine days started with a plenary session at 8.30 am with another at 2.00 pm. These sessions served as an occasion for participants to raise issues of concern for discussion with members of other groups, and added some formality and punctuality to an otherwise informal programme. At the end of each task, groups presented their reports in a plenary session for a final discussion of issues.

The new approach proved highly successful with the Ghanaian participants of the second course. As part of the acceptability testing, the mid-course nominal group exercise showed its eight best features: the range of sectors represented in the course; the group work; the relevance of the field trip; co-operation between participants; the fact that the course is intellectually taxing; the informal

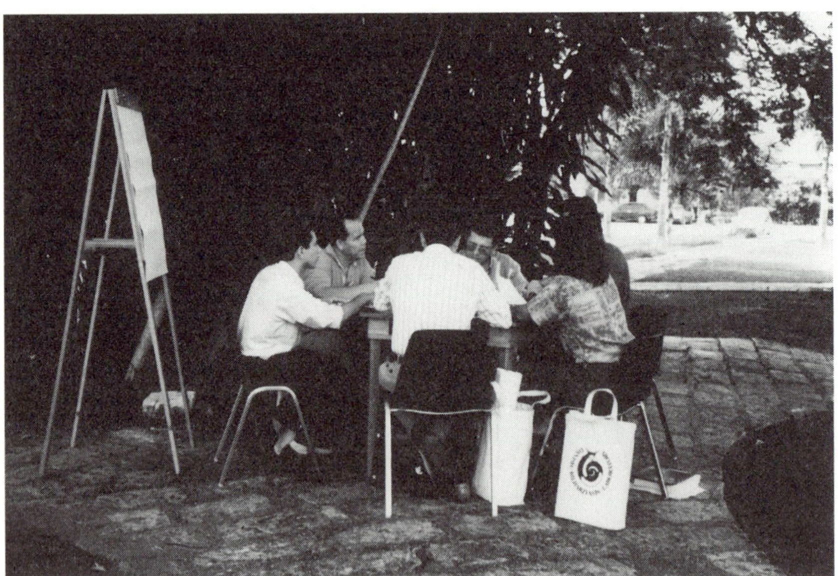

Figure 7.1 Intersectoral group work, El Zamorano, Honduras (photograph by P. Furu).

interaction between participants, course staff and resource personnel; the relevance of the course contents; and the innovative approach of the course.

The course timetable was designed in a way that ensured that the nature of the task and the time allocated to it became progressively more challenging. The initial health impact assessment was the most drawn out task because it covered a lot of new concepts and fields and also served to accustom the groups to the course methodology. The role of the non-expert tutors, who accompanied each group to assist them in group dynamics, became progressively less intense to the point where they were withdrawn completely by the time the groups had started their final task (Figure 7.1).

7.4 The tasks and the *Task Guides*

After the first experience of the consistent problem-based learning approach in Ghana, the course structure was consolidated by a rearrangement of the tasks, the addition of an extra task at the beginning of the course and a refinement and simplification of the *Task Guides*. An external consultant was invited to participate as an observer of the process in the third course in Tanzania and to provide recommendations.

The details of each of the tasks, their aims, learning objectives and conclusions are presented below as they appeared in the report of the Tanzania course, with some minor updates and modifications (WHO 1995).

7.4.1 Task 1 – Constructing a comprehensive development planning framework

Aims

To describe the project planning procedures in the country and to identify mechanisms for coping with otherwise unforeseen effects.

Learning objectives

- to produce an approximate flowchart which indicates the main stages of the project cycle as it is used in the country, the responsibilities of each ministry and the mechanisms for dealing with unexpected and unintended effects (answering the question: what happens currently?);
- to focus on the water resources development sector, identifying its main subsectors;
- to focus further on what the flowchart should look like for key categories of water resources development projects in a given country (answering the question: what should happen ?);
- to emphasize the need for and potential of impact assessment (answering the questions: what could go wrong and what can be done about it ?).

Conclusions

Water resources development generally proceeds in logical stages that include identification, design, pre-feasibility, approval, feasibility, appraisal, redesign, construction and operation. This type of project always has the potential to cause unexpected and unintended impacts that may involve environment, social systems, economic systems and/or health. They require assessment prior to commission in order to anticipate and prevent possible negative impacts and maximize the development potential. They require an intersectoral approach because many sectors are involved in planning and are needed to solve potential problems.

7.4.2 Task 2 – Health impact assessment: a preliminary step

Aims

To determine if an in-depth health impact assessment is necessary and, if so, to convince the government by reasoned argument that an in-depth assessment is needed and that recommendations for health safeguards, mitigation and health improvements are practical.

Learning objectives

- to identify health hazards associated with a given development project;
- to illustrate that development projects can change health risks;
- to illustrate that development plans and operations can be modified to promote health;
- to provide a simple method for assessing health impact and gain first-hand experience of its use;
- to illustrate the types of intervention that may be required to protect and promote health.

Conclusions

There are many health hazards associated with development projects. A rapid assessment of health impacts and opportunities can be made by intersectoral dialogue, study of the project plans and local health data and a brief field visit. Two sequential subtasks are given in Table 7.2. The assessment can be used as a basis for deciding whether a more detailed assessment is needed. Practical health safeguards, mitigation and promotional measures can be recommended.

Table 7.2 Two sequential pages from Task 2

An initial health impact assessment (HIA) procedure

In your discussion of procedures you may have identified the need for screening (i.e. deciding whether a health impact assessment is needed), catagorizing and commissioning further studies. The procedure should consist of the three steps listed below, each under the responsibility of a specialist.

- The proposed new project is probably screened by a non-health specialist who must decide whether a health impact assessment is needed.
- A rapid assessment is then made by an officer with general responsibilities for such assessments. It is designed to spot important health risks and opportunities, and to classify the project according to the probable importance of the impacts.

Classification	*Interpretation*
A	Significant* health impacts, mitigation difficult or requires special budget
B	Significant health impacts, mitigation practical without special budget
C	No significant health impacts

* A health impact is significant if a change in the well-being of a significant number of people can be attributed to the project. In fact, there are health hazards even without the project and we will only consider a change in the health risks attributable to the project.

- Next, a decision is made as to whether a full health impact assessment is necessary. This could be undertaken by a consultant and would be expensive and time-consuming. Additional procedures would then be necessary for selecting an appropriate consultant, writing the terms of reference and appraising the consultant's report. Whether a full assessment is carried out only for category A projects or for both category A and B projects is a matter of national policy and will usually depend on the availability of resources

Question

Why do you think that the decision is made by a non-health specialist?

Discuss this question before proceeding to the next page

Identifying health hazards

The first decision is made by a non-health specialist because this is where the project is instigated. The project officer has to ask the question: *could this project have any health risks?*
The second subtask consists of identifying health hazards. You will be asked to:

- define health hazard and risk
- give examples of a range of health hazards
- classify the health hazards.

Questions

How would you define the terms 'health hazard' and 'health risk'?
Why could it be useful to distinguish between them?

Discuss these two questions before turning to the next page

7.4.3 Task 3 -Technical appraisal of a health impact assessment report

Aims

To learn how to evaluate critically a health impact assessment (HIA) report of a development project made by a health consultant and to recommend acceptance or rejection.

Learning objectives

- to recommend whether to accept or reject the consultant's report and provide an adequate justification of the decision;
- to decide if the report conformed to the TORs and whether the TORs were adequate;
- to determine whether the report had important biases and traps;
- to decide whether the data and their interpretation were sufficient to support the conclusion;
- to decide whether the recommendations made in the report were sufficient to support the conclusion.

Conclusions

Consultants' reports are often inadequate. Care should be taken to provide good TORs, select good consultants and avoid biases and traps. TORs are usually inadequate and there is a need for ideal, generic TORs. This conclusion anticipates Task 5 and confirms the logical structure of the course.

7.4.4 Task 4 – Appraisal of the economic evaluation component of a health impact assessment report

Aims

To appraise the economic evaluation component of an HIA, relating to the recommended options for interventions, in the larger context of the project's overall economic evaluation, and to submit the results to the national authorities responsible for project approval.

Learning objectives

- to learn the language of economic analysis;
- to decide what kind of economic analysis of health impacts had been or could be made;
- to identify what costs and financial benefits had been, or could be, included in the appraisal and assess their validity;

- to assess what health effects had been included;
- to compare net costs of different strategies to manage health outcomes;
- to decide if the assumptions made were realistic.

Conclusions

Economic analysis provides a common language suitable for intersectoral decision-making. The economic evaluation of recommended design changes and interventions must be considered in the context of the overall project evaluation, to ensure that it is proportionate and to identify possible trade-offs. A health economist is a must on the team that is commissioned to carry out this appraisal.

7.4.5 Task 5 – Generic terms of reference for a health impact assessment

Aims

To extrapolate from the experience of the previous two tasks the general rules and requirements for an HIA.

Learning objectives

- to list the components of a generic TOR for an HIA;
- to define the optimal pre-conditions and requirements for an acceptable HIA;
- to specify the expected outputs of the assessment.

Conclusions

Health impact assessment should be a component of environmental impact assessment. Policy and legislation are required to support health impact assessment. Terms of reference should be sufficiently specific to ensure an accurate and detailed assessment report.

7.4.6 Task 6 – Intervention and monitoring: a plan for intersectoral action

Aims

To enable the participants to debate and resolve conflicts that could arise between collaborating ministries regarding responsibilities and financial commitments.

Learning objectives

- to formulate an intersectoral plan of action for the implementation of recommendations concerning health;

- to agree on a Memorandum of Understanding specifying responsibilities, logistics, resource sharing and allocation of funds between the ministry of health and the other ministries involved;
- to agree on plans for project monitoring, including a list of health indicators that should be measured.

Conclusions

Shared responsibilities do need to be allocated and agreed in advance. Otherwise, resources may not be allocated and the project, once operational, may not be sustainable.

7.4.7 Task development

At the beginning of each task, the participants received a formal letter of remit. Depending on the task, the letter came on the official stationery of either the ministry of health, the ministry of the environment or the Environmental Protection Council or equivalent. The letter set the scene for the task, requesting an intersectoral effort to reach some well-defined objectives. The letter further contained terms of reference and defined the outputs and deadline expected from the group work.

The role-play is continued at the time of presentations of task reports. For the third course, some of the resource personnel were invited to assume the role of a high-level official. In Honduras, the group presentations for Tasks 2, 3 and 6 were attended by, respectively, the Director General of Health Services, a high official of the Environmental Protection Council and the Vice-Minister of Health. The formal nature of the occasion was further emphasized by the seating arrangements in the room where plenary sessions took place, by informing participants in advance and by asking them to dress formally.

This real-life approach to the tasks reinforces the seriousness with which groups tackled their assignments. Again, in the Honduras course, the task reports were invariably accompanied by a covering letter responding to the original letter of remit received.

The structure and methodology proposed in the *Task Guides* are illustrated by the example of a page in Table 7.2. One or more concepts were introduced, questions were raised for the group to discuss, and on the next page the answer was given. The groups were instructed not to proceed to the next page until they reached consensus on the question(s) asked.

At the end of the task, groups submitted a written report (on which they received feedback during the course) and presented their findings orally in a plenary session. These occasions were used to help them improve their presenting skills, which were an essential part of the overall exercise.

Field visits are an essential part of the course programme (Figure 7.2) and also serve to break the monotony of the routine group work. The first field visit was

Figure 7.2 Field work is essential – assessing the health risks, Flores Irrigation Scheme, Honduras, June 1996 (photograph by P. Furu).

fully integrated into Task 2 and entailed the collection of on-site information from stakeholders (the health centre, village chief or mayor, local engineers and agricultural extension workers, school headmaster and staff, and representatives of the affected communities themselves) in an area where water resources development was being planned. This trip taught participants to prepare their assignment in such a way as to facilitate collection of a large amount of relevant information within a limited time. The second field trip was more one of observations concerning risks in existing irrigation schemes that were planned and constructed without special attention to health.

The course was set up as a learning experience and not as a workshop. In other words, it is the process leading to the outputs that is the focus of attention rather than the outputs *per se*. Even though after the first one courses have used real project proposals, the results of initial assessments, appraisals of impact statements or intersectoral action plans have not been formally passed on to national authorities. As no ministry or authority commissioned the assignments, it therefore seemed incorrect to pass on the results officially. The generic terms of reference, which do not address any project in particular, are submitted to the national environment authorities for possible inclusion in the overall environmental impact assessment procedures. The non-expert tutors assist in consolidating the outputs of the various groups into one final version during the time the groups are working on Task 6.

7.5 The role of non-expert tutors, local resource personnel and the course director

To prepare for and manage the training course in its present, innovative form requires inputs that are equally unconventional. The human resources used are therefore given special instructions prior to the course, at least a proper briefing at the time they join the course and a debriefing at the time of separating from the course.

The non-expert tutors are individually assigned to each group and their main task is to assist the group in the process of carrying out its tasks. This includes time management, conflict resolution, promoting alternative ways of arriving at a consensus, facilitating consultation of information sources and liaising with the course organizers. As the name indicates, the non-expert tutors are not with the groups for technical consultations.

Two days before the course, the tutors receive an intensive induction, during which the course methodology and their roles are clearly explained. These induction sessions take them through the entire set of course materials, not to highlight the technical issues, but to point out where groups may meet hurdles or get into conflict, and which possible solutions can be considered. The tutors are also instructed to develop a bond with their groups that is characterized by confidence without familiarity. Throughout the course, until Task 6 when they are withdrawn from the groups, tutors meet with the course director at the end of each day to report on progress, problems and incidents and to learn from one another's solutions.

Starting with the second course, it was decided to make use of local experts as resource personnel rather than fly in a group of expatriate experts. Apart from the obvious financial advantage of this change, it was also felt that local experts would be more in touch with the reality of development, health and environment issues in the country and therefore of more practical value to assist the participants. The role of local experts was passive as the courses were geared entirely towards problem-based learning. Local experts were introduced to the groups at the beginning of each relevant task, and the groups were encouraged to consult the experts whenever they could not resolve an issue amongst themselves. Local experts may spend a considerable time unoccupied during their attendance, and this posed a problem during the Ghana course. In the subsequent courses, local experts brought work of their own to do while they were waiting to be consulted.

In addition to the local experts, participants received an individual course package with relevant publications, and a provisional library was set up.

The course director, also referred to as the anchorman, played an overall co-ordinating role in the course. The director's task started with the induction sessions for the non-expert tutors. During the entire course he/she served as a facilitator, chairing plenary sessions, daily meetings of tutors and briefing and debriefing the local experts. Administrative staff took care of course registration, payment of emoluments and the logistical arrangements for the field trips. The course director also maintained contacts with the local authorities, arranging for their participation in sessions when task reports were presented, and with the host institution. In principle, participants channelled their concerns, observations and problems to the course director through the tutors; in practice, the group as a whole rapidly developed cordial relationships and communications were more informal. The final task of the course director was to prepare a report which included an analysis of the evaluation exercises.

The organizers have gradually refined the profile of the ministry staff that should participate in these courses. They should be mid-level, with a good professional training and with a guarantee that they will stay working in their field in the ministry for some time to come. On a personal level, candidates for participation should have a strong interest in working with professionals with different backgrounds and they should be outspoken and good communicators. The international organizers set up a rigorous procedure for selection, which started with the identification of relevant ministries and authorities. These received a formal invitation to submit the names and curricula vitae of candidates whom they wished to nominate for the course. A local selection committee, consisting of the counterpart organizing institution and staff of country offices of WHO and FAO then made a final recommendation to the international organizers. A shortlist of reserve participants was necessary to cater for any last minute withdrawals.

As gender balance was extremely poor in the first two courses, a special effort was made to promote the submission of female nominees by the ministries for the Tanzanian course. Clearly, women are more under-represented in some ministries than in others, and this was reflected in the nominations. As a result, the Tanzania course started off with nine female and twelve male participants. Two of the women were pregnant, however, and after one of them had gone into labour and given birth after the first bumpy field trip, the other thought it advisable to discontinue the course. The two women and the (healthy) baby returned safely to Dar es Salaam, leaving the course with two participants fewer. Needless to say, for the Honduras course the emphasis on balancing female representation was qualified somewhat to non-pregnant women! Seven of the twenty-four participants in Central America were women. They made an important contribution to the success of the course, although in group work they tended towards the secretarial part of the tasks while the men led discussions and took care of oral presentations. In India, however, there was only one woman participant.

7.6 Analysis of the approach and options for follow-up

The training course 'Health Opportunities for Water Resources Development' has evolved into a finely-tuned capacity-building effort which effectively aims at strengthening skills for intersectoral decision-making. The evaluations consistently show a high level of acceptability in terms of methodology, course content and structure. While most participants indicated that the course was more demanding than others they have attended, they also felt it was more beneficial. The products of the tasks invariably showed improved capacity to better understand and solve problems of an intersectoral nature, to arrive at a workable consensus and to present the conclusions of the group in a powerful, concise and convincing manner.

The efficiency of the course greatly improved over time. There was a dramatic difference in its costs when the traditional information transfer approach of the Zimbabwe course was changed into the task-oriented, problem-based learning approach that characterized subsequent courses. In its present form, the course has great potential for replication at a national level, although we believe that ideally the course should be institutionalized to serve a region – the saturation level in the national context is probably quite low.

The organizers have stayed with their original development goals, although there have been frustrations over the fact that no proper follow-up took place in some countries where the course was held. In Zimbabwe, an in-depth health impact assessment of the Mupfure Irrigation Scheme was carried out with the help of a number of course participants. Similarly, policy studies were carried out, also with the help of course participants, to assess to what extent the lack of a conducive policy framework hampers intersectoral collaboration in development planning. The results of these efforts were discussed at a follow-up seminar in Harare in November 1993. The Ghana course was also given some level of follow-up, and a seminar was held in Accra in August 1995.

The possibility of institutionalizing the course has also been investigated and some attractive options exist in principle. However, in the first instance a needs assessment would have to be carried out in the region, linked to a survey that shows to what extent the various ministries are willing to use their limited funds for capacity building to have their staff attend a course of an intersectoral nature.

The current position of the organizers is that the original development objectives (i.e. the development of a balanced, technically sound and sustainable package of course materials) should be met and that any further activities should be part of a next, distinct phase.

The experience of the courses held has also made clear that the methodology applied is suitable for all sorts of problem solving of an intersectoral nature and is not confined to water resources development associated vector-borne disease problems. In fact, the course in Honduras de-emphasized this particular health scope and addressed health in a broader fashion. This reflected the realities of Central America as opposed to countries in Africa. With respect to the development focus, this could also go beyond strict water resources.

The fifth course, in India in 1997 (WHO 1998), was an experiment in this direction. Twenty middle-level managers with backgrounds in health, engineering, irrigation, forestry, environment and public works took part in a course convened at the Water and Land Management Institute, Aurangabad. This course aimed at developing the capacity of each participant, irrespective of sectoral affiliation, to take part adequately in the intersectoral decision-making process of water resources development and to ensure that human health considerations were included. Task 1 was a stage setting exercise and Tasks 2–6 were carried out in the context of existing water resources projects then at planning stage (Som Kamta Amba Irrigation Project in Rajathan, Dholisamel in Gujarat and the Mordhana in Tamil Nadu.) Field trips to the Shivna Takli and to the Sukhna Medium Irrigation Project were also included. From this it is expected that generic forms of reference for HIA will be forwarded to the relevant states. The practical nature of this output generated through intersectoral participation, and with the assistance of non-expert tutors and local experts, is evidenced below.

7.7 Generic terms of reference for health impact assessment of water resources development projects in India

7.7.1 Preamble

The responsibility for human health cuts across all public sectors responsible for development projects. HIA is an integral and indispensable part of EIA. It is of particular importance for large- and medium-scale water resources development (WRD) projects.

In accordance with the policy of the Government of India under the Environment (Protection) Act 1986, EIA is a mandatory State Government procedure for all such projects. The Ministry of Environment and Forests of the Government of India is the central authority for the appraisal of EIA reports and for the endorsement of environmental impact statements.

The purpose of HIA is to identify, evaluate and interpret the hazards and health opportunities of WRD and to propose health safeguards, mitigating measures and health promotional measures that can be incorporated into the WRD project without affecting its overall objectives. The concept of the hidden costs of development due to ill-health supports the need for HIA.

The HIA should be carried out at the feasibility stage of project planning by a multidisciplinary team composed of, preferably national, experts. The outcome of an HIA will focus attention on major health risks, assist in comparing and ranking them and thus help in examining options for possible interventions.

7.7.2 Criteria and preconditions for HIA

At the pre-feasibility stage of any medium- or large-scale water resources development project, the government agency proposing the project (the project proponent) should, in close collaboration with the department responsible for public health, carry out a rapid HIA. This should lead to a decision about whether or not to carry out a full HIA. If affirmative, the project proponent is responsible for the tasks described below, ensuring that the criteria and pre-conditions are met, and starting with the formulation of specific TORs whose generic outline is presented in the next section.

Composition of a multidisciplinary team and selection of a consultant/team leader

The tendering should define the characteristics of the team, in particular which disciplines must be represented by its subject specialist members.

The profile of the consultant/team leader should include the following criteria.

- The consultant should have appropriate qualifications with a broad vision and the capacity to integrate the information from various disciplines.
- The consultant should have sufficient experience in assessing the health impacts of WRD projects.

Timing

The timing of the EIA/HIA should be decided:

- to allow optimal opportunity for the team to collect seasonally fluctuating health data (e.g. seasonal transmission of malaria) if these are not readily available from the health authorities;
- to allow effective communication between the HIA consultant and the consultants working on the EIA and the overall feasibility studies.

The relationship between the team leader and the commissioning authority

- The team must carry out the tasks given in the TORs comprehensively, but should not be limited by the TORs.
- The team leader and the team members will have full independence as far as their tasks are concerned.
- All the information relevant to project and local assistance as and when required will be provided by the commissioning authority.
- The commissioning authority will also facilitate meetings with heads of departments, field visits and health services, etc.
- If an international consultant is hired, a team of local experts must also be involved in the assignment.
- The commissioning authority will submit the HIA report to a formal, independent appraisal procedure, on the basis of which it can be accepted, partly accepted or rejected. If it is partly rejected, the consultant is bound to complete the assignment to conform with the TORs. The commissioning authority will inform the consultant/team leader about the criteria applied in the appraisal procedure at the start of the assignment.

Conditions for financial support

- The funding of an EIA, including an adequate HIA component, should be proportionate to the budget of the feasibility study.
- The government will apply norms and procedures to determine the mode of payment as per agreement.

Other considerations

It is desirable that the consultant submit the draft HIA report to the commissioning authority for circulation among the different government departments concerned, that he/she give serious consideration to the comments and suggestions made by these departments and make amendments to the final report as justified.

The consultant should submit the required number of copies of the final HIA report.

For breach of contract from either side or in case of any dispute action will be taken as per agreement.

7.7.3 Generic terms of reference

Introduction

The introduction should explicitly state the following aspects:

- a short description of the project;
- geographical boundaries of the project;
- existing institutional infrastructure and arrangements (public and private sector);
- sources of relevant data and institutions that can assist in the HIA.

Objectives

The objectives of the HIA should focus on:

- the assessment of health risks and opportunities;
- the suggestion of alternative interventions and mitigating measures required at the appropriate stages, as well as health promotional measures;
- the need to define health safeguards that are technically sound, economically feasible and socially acceptable.

The scope of the HIA

- Geographical and time boundaries for the HIA. If there are geographical boundaries, the scope may be extended beyond the project boundaries to include the catchment area and the area downstream from the project; the

areas from which migrants originate should also be considered. The time horizon for the impact assessment needs to be clearly defined.
- Vulnerable communities (including those formally or informally migrating into the project area).

Activities to be carried out by the consultant

- Careful study of the project documents.
- Collection and analysis of relevant, existing health data from government and non-governmental institutions, complemented where necessary and feasible through field surveys employing appropriate sampling techniques.
- Identification of different health hazards within the geographical boundaries of the project.
- Determination of community, environmental and institutional risk factors.
- Identification of health opportunities.
- Formulation of health risk management measures (alternative interventions/ mitigating measures at different project stages with a justification, including changes in design, in project operation and improvements in the health services). Such measures should pay due regard to community participation, gender issues and social equity.
- Appropriate economic evaluation of alternative risk management measures.
- Provision of indicators in the problem areas for specific diseases needing special attention and in-depth studies.
- Preparation of an HIA report, containing, as the output of the exercise, a health impact statement.

References

Birley, M.H., Bos, R., Engel, C.E. and Furu, P. (1995) Assessing health opportunities: a course on multisectoral planning. *World Health Forum*, **16**, 420–422.

Birley, M.H., Bos, R., Engel, C.E. and Furu, P. (1996) A multisectoral task-based course: Health Opportunities in Water Resources Development. *Education for Health*, **9**, 71–83.

Engel, C.E. (1992) Problem-based learning. *British Journal of Hospital Medicine*, **48**, 325–329.

Tiffen, M. (1989) *Guidelines for the incorporation of health safeguards into irrigation projects through intersectoral cooperation*. Joint WHO/FAO/UNEP Panel of Experts on Environmental Management for Vector Control, WHO/VBC 89.5, WHO, Geneva.

United Nations (1993) *Agenda 21: Programme of Action for Sustainable Development*. UN, New York.

World Commission on Environment and Development (1987) *Our Common Future*. Oxford University Press, Oxford.

World Health Organization (1988) *Report of the eighth meeting of the joint WHO/FAO/UNEP Panel of Experts on Environmental Management for Vector Control (Nairobi, 5–9 September 1988)*. Document VBC/88.4. PEEM Secretariat, Geneva.

World Health Organization (1993) *Health Opportunities in Water Resources Development, with special emphasis on the prevention and control of water associated vector-borne diseases. Report of a training course in Darwendale, Zimbabwe, 7–18 September 1992*. Document WHO/CWS/93.2. PEEM Secretariat, Geneva

World Health Organization (1994) *Health Opportunities in Water Resources Development. Report of a two-week course to promote collaboration between middle-level officials from various ministries for the incorporation of health safeguards and health promotional measures in water resources development, Akosombo, Ghana, 26 January to 11 February 1994*. Document WHO/EOS/94.37. PEEM Secretariat, Geneva.

World Health Organization (1995) *Health Opportunities in Water Resources Development. Report of an 18-day course to promote collaboration between middle-level officials from various ministries for the incorporation of health safeguards and health promotional measures in water resources development, Arusha, Tanzania, 13–30 March 1995*. Document WHO/EOS/95.5. PEEM Secretariat, Geneva.

World Health Organization (1996) *Health Opportunities in Water Resources Development. Report of an 18-day course to promote collaboration between middle-level officials from various ministries for the incorporation of health safeguards and health promotional measures in water resources development, El Zamorano, Honduras, 9–28 June 1996*. Document WHO/EOS/96.17. PEEM Secretariat, Geneva.

World Health Organization (1998) *Report of an 18-day course to promote collaboration between middle-level officials from various ministries for the incorporation of health safeguards and health promotional measures in water resources development*, Aurangabad, India, 10–28 November 1997. *Health Opportunities in Water Resource Development*. Document WHO/EOS/98.9. PEEM Secretariat, Geneva.

Chapter 8

Ord River irrigation area: the effect of dam construction and irrigation on the incidence of Murray Valley encephalitis virus

John S. Mackenzie and Annette K. Broom

8.1 The Ord River irrigation area

The Ord River irrigation area in the north-east of the Kimberley region of Western Australia has been established for over thirty years. After a series of feasibility studies initiated in 1941 to determine the potential for tropical irrigated agriculture, the Western Australian Government became convinced of the viability of an irrigation scheme. Thus in 1958, in collaboration with the Federal Government, stage one of the project was begun. Completed in 1963, it consisted of the construction of the Kununurra diversion dam together with irrigation and associated works, and development of the Kununurra township. By 1966, thirty-one farms irrigated from the diversion dam had been allocated. Stage two, the construction of the Ord River dam to provide a major storage reservoir, Lake Argyle, was completed in 1972.

The Ord River irrigation area is composed of seven major areas comprising approximately 79,000 ha in the Wyndham–East Kimberley shire, with 20,000 ha in the Northern Territory. At present, two of these areas, Ivanhoe Valley (approximately 11,000 ha) to the north of Kununurra and Packsaddle Plain (approximately 2,000 ha) to the south-west of Kununurra, have been developed.

The Ord River dam, 50 km upstream of the diversion dam, provides a water storage volume of 10,760 million m^3, as Lake Argyle, covering a surface area of 740 km^2, and a flood storage volume of 34,655 million m^3 with a surface area of 2,072 km^2. The diversion dam (Figure 8.1) provides a water storage of 98 million m^3 and a storage length, as Lake Kununurra, of 40 km. More than 135 km of channels and over 144 km of drains supply water to the irrigation areas from Lake Kununurra by either gravity flow (to Ivanhoe Plain) or pumping (to Packsaddle Plain). Further details of the two dams and the irrigation areas have been described by Stanley (1972; 1975) and Roberts (1993).

The Kimberley region in the north of Western Australia comprises an area of over 265,000 km^2 and has a tropical, monsoonal climate with distinct 'wet' and 'dry' seasons. It was first settled and has generally remained as cattle country. During the wet season from December to April, rainfall is received erratically

Figure 8.1 The diversion dam on the Ord River, Western Australia.

from cyclonic depressions. Several large rivers in the Kimberley and in the adjacent western area of the Northern Territory, including the Fitzroy, Ord, Victoria and Daly rivers, drain an area of 543,000 km^2 and yield 20 per cent of the total run-off of the continent (Millington, 1975). Flash flooding during the wet season is common, with flows of 30,000 m^3/sec. The average annual rainfall in the Ord River irrigation area is 787 mm. During the dry season in the Kimberley region, which is very dry, water courses tend to dry up, little surface water remains, and seasonal pastures die. Human habitation is sparse in the Kimberley region, with a population of about 28,000. Most people live in one of the six townships (Wyndham, Kununurra and Halls Creek in East Kimberley; Fitzroy Crossing, Derby and Broome in West Kimberley) or in Aboriginal communities.

8.2 Human health – early concerns

The Ord River was the first of the large rivers in tropical Australia to be dammed for irrigation and intensive tropical agriculture. The establishment of large, permanent waterbodies and widespread irrigation would be expected to exert profound changes on the ecology of the area, and it was therefore very important that these changes should be monitored and their effects assessed. Ecological changes brought about by dam construction and irrigation in other tropical countries had been shown to be associated with an increased incidence of human diseases such as malaria, filariasis, schistosomiasis and arboviral diseases (Stanley 1972). Of these, the most important to human health in north-west Australia were arboviruses, and it was anticipated that the Ord River irrigation area would attract increased waterbird populations and provide additional mosquito breeding habitats, which would be conducive to increased arbovirus activity (Stanley 1972; 1975). Indeed the potential problems were expected to become more acute as the population in the area increased with the development of Kununurra township and nearby farming activities, and with increased tourism and mining opportunities.

More than sixty-five arboviruses have been isolated in tropical Australia, but only a few have been implicated in human disease (Mackenzie *et al.* 1994a). These include the flaviviruses Murray Valley encephalitis (MVE), Kunjin, Kokobera, Alfuy, Edge Hill and dengue; and the alphaviruses Ross River, Barmah Forest, and Sindbis (Mackenzie *et al.* 1994a; 1994b). With respect to the Ord River irrigation area, the most important of these viruses is MVE, the major cause of Australian encephalitis. MVE virus has a natural biocenose between waterbirds, particularly members of the order Ciconiiformes, and mosquitoes, particularly the fresh-water breeding species, *Culex annulirostris*. MVE virus is a member of the Japanese encephalitis serological complex of flaviviruses, and is more closely related to Japanese encephalitis virus than are the other Australian members of the complex (Kunjin, Kokobera, Alfuy and Stratford viruses).

Table 8.1 Australian encephalitis: number of clinical cases 1951–1995

Year	*Australian State or Territory*[a]						*Total*
	WA	*NT*	*QLD*	*VIC*	*NSW*	*SA*	
1951	–	–	–	34	10	1	45
1956	–	–	–	3	–	–	3
1969	1	–	–	–	–	–	1
1971	–	–	1	–	1	–	2
1974	1	5	10	27	5	10	58
1978	5[b]	–	–	–	–	–	5
1979	1	–	–	–	–	–	1
1981	7	1	2	–	–	–	10
1984	2	–	–	1[b]	–	–	3
1986	1	–	–	–	–	–	1
1987	–	1	–	–	–	–	1
1988	–	3	–	–	–	–	3
1989	1	–	–	–	–	–	1
1990	1	–	–	–	–	–	1
1991	2[b]	2	1	–	–	–	5
1993	9	6	–	–	–	–	15
1994	–	–	1	–	–	–	1
1995	1[b]	–	–	–	–	–	1

[a] WA = Western Australia; NT = Northern Territory; QLD = Queensland; VIC = Victoria; NSW = New South Wales; SA = South Australia.

[b] One case in each of these years was caused by Kunjin virus. All other cases since 1978 were due to Murray Valley encephalitis virus.

Little was known about MVE virus, its vertebrate hosts or its vectors before the establishment of the Ord River irrigation area. Early serological studies by Stanley and Choo (1961; 1964) on human sera collected in 1960 from Halls Creek in East Kimberley and Derby in West Kimberley had demonstrated that the virus was circulating in these areas. However, no clinical cases of encephalitis had been reported, which may have been due to the small human population in the region prior to 1960, to a lack of awareness by clinicians, to low virus carriage rates in mosquitoes, or to a combination of these factors. Similarly, no cases of encephalitis had been reported in the Northern Territory. The first clinical case of Murray Valley encephalitis (now known as Australian encephalitis) occurred in 1969 (Table 8.1), a fatal case that was acquired by a tourist south of the Ord River irrigation area (Cook *et al.* 1970). Only limited information was available on the mosquito species prevalent in the Ord River area before 1972, although *Culex annulirostris*, believed to be the major vector for MVE virus from studies carried out by Doherty and colleagues in north Queensland (Doherty *et al.* 1963), was found to be present (H. Paterson, personal communication to Stanley 1972), and was the dominant species (H. Paterson, personal communication to Stanley 1975).

Thus prior to the completion of stage one of the Ord River irrigation area, serological evidence had been obtained to demonstrate that MVE virus caused subclinical human infections, but no clinical cases had been reported. Between the completion of stage one and stage two, the first clinical case of encephalitis was reported, and limited information on the mosquito fauna was obtained but without details of mosquito numbers or population dynamics.

8.3 Studies on Murray Valley encephalitis from 1972

8.3.1 Early studies, 1972–1976

A series of investigations on the ecology of MVE virus in the Ord River irrigation area and on the effect of the completion of the Ord River dam were initiated by Stanley and colleagues in 1972. The major components comprised: regular mosquito collections obtained just before and immediately after the wet season to determine the number and proportion of each species at different sites, and for isolation of viruses; serological studies of animals and birds to investigate their roles as possible vertebrate or reservoir hosts; and serological studies of the human population, both Caucasian and Aboriginal, to determine subclinical infection rates and to assess potential risks.

These studies yielded a number of important findings which have provided the basis for much of our knowledge of MVE ecology in north-western Australia. The major findings were as follows.

- Mosquitoes. Using live bait traps to collect mosquitoes, it appeared that there had been a significant increase in mosquito numbers since the construction of the diversion dam (Stanley 1979). Most of the species collected in the bait traps were those associated with permanent and semipermanent fresh water breeding sites, and the dominant species was *Culex annulirostris*, which accounted for over 70 per cent of the collections (Liehne *et al.* 1976a; Stanley 1979). Thus the major vector species for MVE virus was shown to be abundant in the Ord River irrigation area. The major mosquito breeding areas were in swampland adjacent to the diversion dam. Little breeding activity was found in the irrigation area probably due to the excessive use of insecticides applied by aerial spraying for controlling insect pests on cotton crops. However, cotton was discontinued as a crop in 1975 , and an increased number of mosquitoes began to appear in 1976.

- Viruses. Pools of mosquitoes were processed for virus isolation by intracerebral inoculation of macerated mosquito supernatants into suckling mice. A total of 195 strains of 16 arboviruses were isolated from 1075 pools, of which 29 were identified as MVE virus and 21 as Kunjin virus. The majority of the isolates were made from *Culex annulirostris* (153 of 195 isolations), including 28 of 29 identified as MVE. Thus the overall virus isolation rate was high (18 per cent). For MVE virus from *Culex annulirostris*, 3.5 per cent of pools yielded virus at an approximate rate of 1 infected mosquito per 1459 uninfected mosquitoes (Liehne *et al.* 1976b; 1981).
- Serological studies of animals and birds. All the early serological investigations employed the haemagglutination-inhibition (HI) assay. Cattle sera obtained from the Ord River irrigation area exhibited a high incidence of antibody to MVE virus (80 per cent positive), but the incidence declined to 37 per cent positivity in sera collected elsewhere in the Kimberley region (Liehne *et al.* 1976c). A very significant increase in the incidence of antibody to MVE was observed in cattle between 1972 and 1975 in the irrigation area and nearby cattle properties, with increases ranging from between 22 and 36 per cent to between 75 and 90 per cent (Stanley 1979). While the establishment of the irrigation area and the completion of the Ord River dam were undoubtedly responsible for some of this increase, it is probable that the very heavy 'wet' season rainfall in 1973–74 also contributed.

  Of 335 sera collected from 31 avian species, 195 were found to have antibody to MVE virus. Although only a few species were sampled in moderate or large numbers, it was interesting to note that the incidence of antibody was similar between waterbirds and non-waterbirds (56 and 59 per cent, respectively), and between different avian orders: Ciconiiformes (herons, egrets), 62 per cent; Anseriformes (ducks, grebes), 55 per cent; and Psittaciformes (parrots), 56 per cent, (Liehne *et al.* 1976c).
- Human serological studies. A total of 441 human sera were collected in the Ord River area, of which 293 were from Caucasians and 148 from Aboriginals. A very high incidence of MVE antibodies was observed in the Aboriginal population, with 96 per cent of adults and 77 per cent of children exhibiting antibodies. In the Caucasian population, the incidence of MVE

virus antibodies was 53 per cent in adults and 24 per cent in children, but the length of residence in the Ord River area was an important determinant, with those who had lived in the area fewer than three years having a lower incidence (26 per cent) than those who had lived in the area for more than three years (64 per cent) (Liehne *et al.* 1976c).

Thus these early results demonstrated that the mosquito density and bird numbers had increased since the establishment of the Ord River irrigation project, particularly around the diversion dam and Lake Kununurra, that the major mosquito vector of MVE virus was the predominant species *Culex annulirostris*, and that MVE virus was actively circulating in the area. However, the serological results must be treated with caution as the HI test cannot differentiate clearly between MVE and Kunjin viruses, and therefore a number of seroconversions may have been due to infection with the latter. Nevertheless, the results suggested that MVE virus may have become enzootic in the Ord River irrigation area. A single case of Australian encephalitis occurred in Kununurra in 1974; this was the last case of the 1974 epidemic that affected all Australian mainland states (Table 8.1). The first cases to be reported in the Northern Territory also occurred during the 1974 epidemic.

8.3.2 Studies carried out between 1977 and 1995

The early studies between 1972 and 1976 laid the foundation for the more detailed investigations of MVE virus ecology in north-western Australia that have been undertaken over the past twenty years. These investigations became increasingly important as cases of Australian encephalitis became more frequent, particularly with respect to surveillance methodology to enable early warnings to be given of impending epidemic activity and to understand the spread and possible persistence of the virus. In addition, the apparently ideal conditions for arboviral ecology in the Ord River irrigation area have made it essential to monitor for possible incursant mosquito vector species and viruses that could potentially become established in the region.

Improved methods for mosquito collection, virus isolation, and antibody detection have been introduced over the past twenty years, which have allowed a more accurate picture to emerge of the ecology of MVE virus and a more effective surveillance system to be established to provide an early warning of increased virus activity. Human cases of Australian encephalitis, surveillance for virus activity, virus isolations, factors affecting mosquito populations, and virus spread and persistence are discussed below.

Human encephalitis cases

Increasing numbers of Australian encephalitis cases have occurred in Western Australia and the Northern Territory since 1977 (Mackenzie and Broom 1995; Mackenzie *et al.* 1993a; Smith *et al.* 1993). Indeed the majority of cases reported in Australia since 1977, thirty of forty-eight cases, have been from Western Australia, with a further thirteen cases from the Northern Territory. It is also interesting to note that the first confirmed case of encephalitis due to Kunjin virus occurred in Western Australia in 1978, and three additional cases have been diagnosed since, two from Western Australia in 1991 and 1995, and one in Victoria in 1984 (Table 8.1).

Most of the cases of Australian encephalitis in Western Australia have occurred in areas distant from the Ord River irrigation area. Of particular significance was the spread of MVE virus from the Kimberley area south to the Pilbara and Gascoyne regions causing one case of encephalitis in 1978 and three cases in 1981. It is hypothesized that movement of virus to the Pilbara region in 1978 was due to an increase in viral activity in the West Kimberley area following heavy rainfall and flooding, and that with subsequent extensive cyclonic rainfall in the Pilbara region, viraemic waterbirds moved south down the narrow coastal strip, introducing the virus into Pilbara (Stanley 1979). It is probable that a similar mechanism may have occurred in 1981. Although there has been evidence (see next section), of MVE virus activity in the Pilbara region in recent years, there have been no further cases.

Analysis of the cases of Australian encephalitis has indicated that Aboriginal infants, particularly male infants, are most at risk of fatal or severe disease (Mackenzie *et al.* 1993a). However, tourists and visitors to the Kimberley region (and Northern Territory) have also been shown to have an increased risk of disease.

Sentinel chicken surveillance

Following the 1978 outbreak of Australian encephalitis, a number of sentinel chicken flocks were established in the Kimberley area. Six flocks had been established by 1981 and the number rose to twenty-four flocks in twenty-two regional centres in the Kimberley, Pilbara and Gascoyne regions by 1989 (Broom *et al.* 1989; Mackenzie *et al.* 1992; 1994c). Each flock contains twelve chickens which are bled at two weekly intervals between November and June, the period of increased risk of virus transmission, and monthly at other times. The sera are then assayed for antibody to MVE and Kunjin viruses in our laboratory in Perth to provide an early warning system of increased virus activity. Initially sera were tested by HI for the presence of antibody, and positive sera were then subjected to neutralization assay to determine the identity of the infecting virus. A more rapid enzyme-linked immunosorbent assay (ELISA) was introduced in 1986 (Broom *et al.* 1987), and more recently a competitive ELISA using specific monoclonal antibodies to identify the virus is being used (Hall *et al.* 1992; 1995). Sentinel chicken flocks were also established in 1992 in the Northern Territory to monitor MVE activity (Aldred *et al.* 1992).

The sentinel chicken programme has clearly shown that MVE virus is enzootic in several areas of the Kimberley region, particularly in the Ord River area at

Kununurra. Seroconversions in sentinel chickens occur every year during the latter half of the wet season in Kununurra; indeed, occasional seroconversions have been recorded in every month of the year. Elsewhere in the Kimberley region, seroconversions occur in most years towards the end of the wet season at all sites monitored, but the overall frequency tends to be less than that observed in Kununurra, except when flooding is extensive and widespread. Until about 1990, most seroconversions in sentinel chickens in the Pilbara region were due to infections with Kunjin virus, but over the next three years seroconversions to MVE virus showed a significant increase in incidence, suggesting that virus movement from the Kimberley region may be occurring more often. Since 1993, however, Kunjin virus activity has once again become more prevalent in the Pilbara area.

Mosquito collections

Continuing studies in 1976 and 1977 in the Ord River area using bait traps showed that while *Culex annulirostris* continued to dominate the mosquito fauna of the area, other species such as *Coquillettidia xanthogaster*, *Mansonia uniformis* and *Anopheles bancroftii* increased in number following stabilization of the margins of Lake Kununurra and the prolific growth of aquatic plant species (Wright 1981). Studies in the West Kimberley area in 1977 in the Derby area also found that *Culex annulirostris* was the dominant mosquito species (Wright *et al.* 1981).

A major advance in mosquito trapping in the north of Western Australia was the introduction of the EVS-CO_2 light trap in 1978, which replaced the use of bait traps after 1979. This resulted in a ninefold increase in the number of mosquitoes being collected, and a significant increase in the species diversity, although *Culex annulirostris* remained the dominant species (Stanley 1979).

Annual mosquito collections have continued to be undertaken in the Ord River area and at other sites in the Kimberley region since 1978, particularly at the end of the wet season although also at other times if unusual environmental conditions such as cyclones or early wet season flooding have occurred. With the stabilization of Lakes Argyle and Kununurra and of the area under irrigation, the results obtained have provided a clearer association between environmental conditions, mosquito numbers and virus activity (see below). Although the mosquito density, and thus the number collected, is always relatively high in the Ord River area, heavy wet season rainfall and flooding result in a significant increase in the mosquito density. In other areas of the Kimberley, a similar pattern has emerged but the increase in the mosquito density is often more marked than in the Ord River area, and the proportion of different mosquito species tends to vary considerably. Nevertheless, regardless of the study area, *Culex annulirostris* dominates after widespread heavy rainfall and flooding, but if the rainfall is more localized, other floodplain breeding species such as *Aedes normanensis* may dominate initially (e.g. Broom *et al.* 1992).

Virus isolations

Mosquito collections obtained during most field trips to the north-west of Western Australia have been processed for virus isolation. Until 1985, virus isolation was undertaken by intracerebral inoculation of suckling mice, but this was then replaced by cell culture using C6/36 mosquito, PSEK, BHK and Vero cells. The use of cell culture has significantly reduced the overall virus isolation rate by largely excluding arboviruses, rhabdoviruses and most bunyaviruses, but is as effective as suckling mice for the isolation of flaviviruses and alphaviruses.

MVE virus has been isolated every year that significant numbers of adult mosquitoes have been processed except 1983 (Broom *et al.* 1989; Broom *et al.* 1992; Mackenzie *et al.* 1994c). Isolations of MVE, Kunjin and other flaviviruses are shown in Table 8.2. There was a strong correlation between the number of virus isolates in any given year and the prevailing environmental conditions. Thus those years with a heavy, above average wet season rainfall and subsequent widespread flooding yielded large numbers of virus isolates (1981, 1991, 1993) compared with years with average or below average rainfall and with only localized flooding. Although most MVE virus isolates were obtained from *Culex annulirostris* mosquitoes, occasional isolates were also obtained from a variety of other species, including *Culex quinquefasciatus*, *Culex palpalis*, *Aedes normanensis*, *Aedes pseudonormanensis*, *Aedes eidvoldensis*, *Aedes tremulus*, *Anopheles annulipes*, *Anopheles bancroftii*, *Anopheles amictus* and *Mansonia uniformis* (cited in Mackenzie *et al.* 1994b; Mackenzie and Broom 1995), although the role of these species in natural transmission cycles has still to be determined.

Virus carriage rates in *Culex annulirostris* mosquitoes are shown in Table 8.3 for the Ord River area (Kununurra–Wyndham) and Balgo and Billiluna in south-east Kimberley. Very high mosquito infection rates were observed in those years with above average rainfall.

Virus spread and persistence

Stanley (1979) suggested that viraemic waterbirds, which are often nomadic, may generate epidemic activity of MVE in south-east Australia and in the Pilbara region. In an attempt to understand the genesis of epidemic activity better, our laboratory initiated a long-term study in the arid south-east Kimberley area at Billiluna and Balgo, two Aboriginal communities on the northern edge of the Great Sandy Desert. Occasional cases of Australian encephalitis had occurred in both communities (1978, 1981). The studies have clearly shown that MVE virus activity only occurs following very heavy, widespread rainfall both locally and in the catchment area of the nearby watercourse, Sturt Creek, which results in extensive flooding across its floodplain (Broom *et al.* 1992). Localized flooding is insufficient to generate virus activity. Two possible explanations can be proposed to account for the reappearance of MVE virus activity when environmental conditions are suitable: either virus can be reintroduced into the area by viraemic

Table 8.2 Flavivirus isolations from the Kimberley region, 1972–1993[a]

Virus	*1972–1976*	*1977*	*1978*	*1979*	*1980*	*1981*	*1982*	*1983*	*1984*	*1986*	*1989*	*1990*	*1991*	*1992*	*1993*
MVE	29	26	1	–	1	95	1	–	4	1	5	1	46	2	244
Kunjin	21	1	–	2	–	–	4	1	1	1	3	–	9	–	–
Kokobera	2	–	–	–	–	–	–	–	1	4	–	–	–	–	–
Alfuy	–	–	–	–	–	–	–	–	–	–	–	–	2	–	–

[a] Adult mosquitoes were not collected or not processed in 1985, 1987 or 1988

Table 8.3 Murray Valley encephalitis and Kunjin virus infection rates from *Culex annulirostris* mosquitoes in the East Kimberley region

Year	*Location of collection*	*No. mosquitoes processed*	*No. virus isolations*	*Infection rate*	*Human cases*[a]
1972–1976	Kununurra	40,850	28	1:1459	0(1)
1978	Balgo	1,550	1	1:1550	1(4)
1981	Balgo/Billiluna	9,910	87	1:114	2(5)
1982	Kununurra	17,863	4	1:4466	0
1983	Kununurra	6,811	1	1:6811	0
1984	Kununurra	17,375	3	1:5792	0(2)
1986	Kununurra	9,543	2	1:4772	0(1)
1989	Kununurra	12,854	2	1:6427	1
	Billiluna	2,848	4	1:712	0(1)
1990	Kununurra	15,152	1	1:15,152	0(1)
1991	Balgo/Billiluna	4,410	13	1:339	1(1)
	Kununurra	1,716	5	1:343	1(1)
1993	Kununurra	11,015	47	1:234	2(7)
	Billiluna	10,959	114	1:96	1(8)

[a] Bracketed figures = human cases elsewhere in northern Western Australia.

waterbirds arriving from enzootic areas further north; or virus may be reactivated from vertically-transmitted virus in floodwater *Aedes* species. Desiccation-resistant eggs of these mosquitoes may lay around in cracks and crevices for months, and dramatic hatchings after major rainfall are often responsible for mosquito plagues some eight to ten days later.

MVE virus has recently been isolated from male *Aedes tremulus* mosquitoes trapped at Halls Creek in the East Kimberley region, approximately 150 km north of Billiluna and 180 km south-west of Kununurra. The mosquitoes were trapped shortly after the first heavy wet season rains following nine months of drought. Thus this is the first field evidence for vertical transmission of MVE. However, MVE virus was also isolated from blood engorged *Culex annulirostris* mosquitoes trapped at another site in the Halls Creek area on the same night. As there was insufficient time for more than one vector breeding cycle, it is almost certain that the *Culex annulirostris* mosquitoes had been infected by a viraemic vertebrate, possibly a waterbird. Thus the results from these investigations strongly point to two mechanisms for the initiation of virus activity in arid areas of the Kimberley: introduction of virus by viraemic vertebrate hosts; and reactivation of vertically-transmitted virus in desiccation-resistant eggs of some *Aedes* species (Broom *et al.* 1995).

8.4 Discussion and conclusions

We can only speculate on the effect of the establishment of the Ord River Irrigation Area on MVE virus activity in the region. Although seroepidemiological studies on sera collected from Halls Creek in the East Kimberley region and Derby in West Kimberley in 1960 had demonstrated that subclinical infections with both MVE and Kunjin viruses had occurred in the human population (Stanley and Choo, 1961; 1964), there had been no reported cases of Australian encephalitis in Western Australia or in the Northern Territory. Unfortunately no baseline studies were undertaken on either mosquito densities or virus incidence before the completion of stage one of the irrigation project; indeed no studies were initiated until completion of stage two, the construction of the Ord River dam. While the Ord River irrigation area undoubtedly had enormous and profound effects on the ecology of the region, most of the evidence for increases in mosquito densities and waterbird populations is circumstantial.

The climate in the Kimberley and adjacent areas of the Northern Territory comprises a relatively short (four month) monsoonal wet season during which heavy rainfall events occur and the major rivers extend across vast floodplains, and a very dry 'dry' season during which most of the country becomes arid and, in the latter half, even large rivers cease to flow. Results from studies at various locations, such as Billiluna and Halls Creek, suggest that MVE virus is occasionally epizootic in many arid areas of the Kimberley. It is probable, therefore, that the area in which the Ord River irrigation area was established was similar and, consequently, that prior to the irrigation scheme being implemented, MVE was also epizootic.

Since 1972, our studies in the Ord River irrigation area and elsewhere in the Kimberley region on virus isolations from mosquitoes, on serological investigations of humans, animals and sentinel chickens, and on human cases of Australian encephalitis, have clearly shown that MVE virus is now enzootic in the Ord River area and probably in other foci such as the Derby and Broome areas of the West Kimberley region. Elsewhere, in arid areas of the Kimberley and in the Pilbara, MVE virus is epizootic and virus activity is probably initiated either by virus reactivation from desiccation-resistant mosquito eggs or by introduction through viraemic vertebrate hosts. The situation in the Northern Territory is less clear as insufficient data have been accumulated. However, it is probable that MVE is enzootic in the wetlands in the north of the Northern Territory, but epizootic in the more arid areas further south extending east from the Kimberley border. Since 1978 there has been a substantial increase in the number of cases of Australian encephalitis throughout the Kimberley and Northern Territory that cannot be ascribed to either an increase in population or a heightened awareness among clinicians.

Thus, although based largely on circumstantial evidence, we believe that the Ord River Irrigation Area has had a profound effect on MVE virus activity and indeed has resulted in the virus becoming enzootic in the area. We also believe that this large, stable enzootic focus has provided the source for regular epizootic incursions to other areas of the Kimberley and adjacent arid areas of the Northern Territory, and to the Pilbara, and has probably established smaller enzootic foci in the West Kimberley. As virus can persist in desiccation-resistant mosquito eggs, it is probable that most areas of the Kimberley and adjacent areas of the Northern Territory, and possibly parts of the Pilbara, have been 'seeded' with virus which could result in epizootic activity when appropriate environmental conditions occur.

Our conclusions could have important health implications as the population in north-western Australia increases through intensive agriculture, mining, service industries and tourism and, in the longer term, through possible effects of climate change (Mackenzie *et al.* 1993b; Lindsay and Mackenzie 1997). Furthermore, increased virus activity could be exacerbated as new irrigation areas are developed in the Wyndham–East Kimberley shire and the adjacent part of the Northern Territory.

Finally, there is little doubt that the profound ecological changes resulting from the establishment of the Ord River irrigation area have provided ideal conditions for increased arboviral activity. These conditions are also suitable for other exotic arboviruses, such as Japanese encephalitis and chikungunya viruses, and exotic mosquito vectors, such as *Aedes albopictus*. Indeed an unusual strain of MVE has been isolated from the Ord River area, which was believed to have been introduced from the Indonesian archipelago (Mackenzie *et al.* 1991). Furthermore, the recent incursion of Japanese encephalitis virus into islands in the Torres Strait and Cape York, and its possible enzootic presence in the south of Papua New Guinea, provide additional cause for concern. It is therefore essential that monitoring and surveillance of mosquitoes and arboviruses is continued so that exotic virus or vector incursions can be rapidly detected.

Acknowledgments

We would like to thank our many colleagues who have contributed to these studies of MVE virus activity in the north-west of Western Australia. We would also like to acknowledge the support of the Health Department of Western Australia and the National Health and Medical Research Council, and the Commonwealth Department of Health.

References

Aldred, J, Broom, A K, Hueston, L and Mackenzie, J S (1992) Australian encephalitis: sentinel chicken surveillance programme – serological results for March 1992. *Communicable Diseases Intelligence (Australia)* **16**, 169.

Broom, A K, Charlick, J, Richards, S J and Mackenzie, J S (1987) An enzyme-linked immunosorbent assay for detection of flavivirus antibodies in chicken sera. *Journal of Virological Methods*. **15**, 1–9.

Broom, A K, Lindsay, M D A, Johansen, C A, Wright, A E and Mackenzie, J S (1995) Two possible mechanisms for survival and initiation of Murray Valley encephalitis virus activity in the Kimberley region of Western Australia. *American Journal of Tropical Medicine and Hygiene*. **53**, 95–99.

Broom, A K, Lindsay, M D, Wright, A E and Mackenzie, J S (1992) Arbovirus activity in a remote community in the south-east Kimberley. *Arbovirus Research in Australia.* **6**, 262–266.

Broom, A K, Mackenzie, J S, Lindsay M D and Wright, A E (1989) Epidemiology of MVE and Kunjin viruses in Western Australia. *Arbovirus Research in Australia.* 5, 14–18.

Cook, I, Allan, B C, Horsfall, W R and Flanagan, J E (1970) A fatal case of Murray Valley encephalitis. *Medical Journal of Australia.* **1**, 1110–1112.

Doherty, R L, Carley, J G, Mackerras, M J and Marks, E N (1963) Studies of arthropod-borne virus infections in Queensland. III. Isolation and characterisation of virus strains from wild-caught mosquitoes in north Queensland. *Australian Journal of Experimental Biology and Medical Science.* **41**, 17–30.

Hall, R A, Broom A K, Hartnett, A, Howard, M J and Mackenzie, J S (1992) Specific monitoring of flavivirus infections in man and sentinel animals by epitope blocking ELISA. *Arbovirus Research in Australia.* **6**, 155–157.

Hall, R A, Broom A K, Hartnett, A C, Howard, M J and Mackenzie, J S (1995) Immunodominant epitopes on the NS1 protein of MVE and KUN viruses serve as targets for a blocking ELISA to detect virus-specific antibodies in sentinel animal serum. *Journal of Virological Methods.* **51**, 201–210.

Hanna, J N, Ritchie, S A, Phillips, D A, Shield, J, Bailey, M C, Mackenzie, J S, Poidinger, M, McCall, B J and Mills, P J (1996) An outbreak of Japanese encephalitis in the Torres Strait, Australia, 1995. *Medical Journal of Australia.* **165**, 256–260.

Liehne, P F S, Anderson, S, Stanley, N F, Liehne, C G, Wright, A E, Chan, K H, Leivers, S, Britten D K and Hamilton, N P (1981) Isolation of Murray Valley encephalitis virus and other arboviruses in the Ord River Valley, 1972–1976. *Australian Journal of Experimental Biology and Medical Science.* **59**, 347–356.

Liehne, C G, Leivers, S, Stanley, N F, Alpers, M P, Paul, S, Liehne, P F S and Chan, K H (1976b) Ord River arboviruses – isolations from mosquitoes. *Australian Journal of Experimental Biology and Medical Science.* **54**, 499–504.

Liehne, P F S, Stanley, N F, Alpers, M P and Liehne, C G (1976a) Ord River arboviruses – the study site and mosquitoes. *Australian Journal of Experimental Biology and Medical Science.* **54**, 487–497.

Liehne, C G, Stanley, N F, Alpers, M P, Paul, S, Liehne, P F S and Chan, K H (1976c) Ord River arboviruses – serological epidemiology. *Australian Journal of Experimental Biology and Medical Science.* **54**, 505–512

Lindsay, M and Mackenzie, J (1997) Vector-borne viral diseases and climate change in the Australian region: major concerns and the public health response. In: *Climate Change and Human Health in the Asia-Pacific Region* (P Curson, C Guest, and E Jackson, eds.). Australian Medical Association and Greenpeace International, pp. 47–62.

Mackenzie, J S and Broom A K (1995) Australian X disease, Murray Valley encephalitis, and the French connection. *Veterinary Microbiology.* **46**, 79–90.

Mackenzie, J S, Broom, A K, Aldred, J, Hueston, L and Cunningham, A L (1992) Australian encephalitis: sentinel chicken surveillance programme. *Communicable Diseases Intelligence (Australia).* **16**, 55–57.

Mackenzie, J S, Coelen, R J, Lawson, M A, Sammels, L, Howard, M, Hall, R A and Broom, A K (1991) Molecular approaches to the study of the epidemiology of two Australian flaviviruses: Murray Valley encephalitis and Kunjin viruses. *Proceedings of the Australian Physiological and Pharmacological Society.* **21**, 121–131

Mackenzie, J S, Coelen, R J, Sellner, L, Broom, A K, Lindsay, M D, Hall, R A and Oliveira, N (1994c) Surveillance of mosquito-borne viral diseases: a brief overview and experiences in Western Australia. In: *Rapid Methods and Automation in Microbiology and Immunology* (R C Spencer, E P Wright and S W B Newsom, eds). Intercept, Andover, UK, pp. 191–203.

Mackenzie, J S, Lindsay, M D and Broom, A K (1993b) Climate changes and vector-borne disease: potential consequences for human health. In: *Health in the Greenhouse: The Medical and Environmental Health Effects of Global Climate Change* (C E Ewan, E A Bryant, G D Calvert, J A Garrick, eds). Australian Government Publishing Service, Canberra, pp. 229–234.

Mackenzie, J S, Lindsay, M D, Coelen, R J, Broom A K, Hall, R A and Smith, D W (1994b) Arboviruses causing human disease in the Australasian zoogeographic region. *Archives of Virology*. **136**, 447–467.

Mackenzie, J S, Poidinger, M, Phillips, D, Johansen, C, Hall, R A, Hanna, J, Ritchie, S, Shield, J and Graham, R (1997) Emergence of Japanese encephalitis virus in the Australian region. In: *Factors in the Emergence of Arbovirus Diseases* (J F Saluzzo and B Dodet, eds.). Elsevier, Paris. pp. 191–201.

Mackenzie, J S, Smith, D W, Broom, A K and Bucens, M R (1993a) Australian Encephalitis in Western Australia, 1978–1991. *Medical Journal of Australia*. **158**, 591–595.

Mackenzie, J S, Smith, D W, Ellis, T M, Lindsay, M D, Broom, A K, Coelen, R J and Hall, R A (1994a) Human and animal arboviral diseases in Australia. In: *Recent Advances in Microbiology*, Vol. 2 (G L Gilbert, ed.). Australian Society for Microbiology, Melbourne, pp. 1–91.

Millington, A J (1975) Agricultural implications of the Ord River Dam. In: *Man-Made Lakes and Human Health* (N F Stanley and M P Alpers, eds). Academic Press, London, pp. 113–135.

Roberts, F (1993) *Ord River Irrigation Area* (3rd ed.). Western Australian Department of Agriculture, Perth.

Smith, D, Mackenzie, J, Broom, A, Fisher, D, Williams, M, Burrow, J and Currie, B (1993) Preliminary report of Australian encephalitis in Western Australia and the Northern Territory, 1993. *Communicable Diseases Intelligence (Australia)*. **17**, 209–210.

Stanley, N F (1972) Ord River ecology. *Search*. **3**, 7–12.

Stanley, N F (1975) The Ord River Dam of tropical Australia. In: *Man-Made Lakes and Human Health* (N F Stanley and M P Alpers, eds). Academic Press, London, pp. 103–112.

Stanley, N F (1979) Problems related to the epidemiology of Murray Valley encephalitis and Kunjin viruses created by development in north-west Australia. *Arbovirus Research in Australia*. **2**, 41–46.

Stanley, N F and Choo, S B (1961) Serological epidemiology of arbor viruses in Western Australia. *Medical Journal of Australia*. **2**, 781–783.

Stanley, N F and Choo, S B (1964) Studies of arboviruses in Western Australia. *Bulletin of the World Health Organization*. **30**, 221–226.

Wright, A E (1981) Ord River arboviruses – mosquito captures during 1976/77. *Journal of Australian Entomological Society*. **20**, 47–57.

Wright, A E, Anderson, S, Stanley, N F, Liehne, P F S and Britten, D K (1981) A preliminary investigation of the ecology of arboviruses in the Derby area of the Kimberley region, Western Australia. *Australian Journal of Experimental Biology and Medical Science*. **59**, 357–367.

Chapter 9

Safe management of the Ross River dam, Northern Australia, in relation to recreation and health

Brian Kay, Mark Hearnden and Marina Hurley

9.1 Background

Compared with some of the dams in the world, the Ross River dam on the outskirts of the twin cities of Townsville and Thuringowa is just a baby. In 1973, it was constructed largely as an earth-rock wall with a concrete spillway height of 34.6 m to inundate approximately 3,420 ha to an average depth of 2.9 m. In 1987–88, the concrete spillway was raised to 38.2 m, increasing the storage capacity from 72,000 to 236,000 ml at full supply level.

This reservoir is interesting for several reasons. First, it is in tropical Australia where mosquitoes are active all year round and malaria and filariasis used to be endemic. Second, it skirts the major population centre of north Queensland, with approximately 130,000 residents, some of whom live within sight of the man-made lake. Third, because the reservoir is shallow, now covering a maximum of 14,164 ha, dry season shrinkage of the lake results in great expanses of swampy areas eminently suitable as waterbird and mosquito habitat.

Apart from the abundant avifauna, wild kangaroos and wallabies, dogs, pigs, foxes, dingoes, possums and camels are intermixed with cattle and horses on agistment. Thus the situation would seem ideal for arbovirus transmission as many of these are recognized vertebrate hosts of Australia's medically important Murray Valley encephalitis, and Ross River, Kunjin and Barmah Forest viruses. These viruses survive in silent sylvan cycles involving mosquitoes and wild and domestic vertebrates and occasionally spill over into the human population to cause epidemics. Furthermore during summer, north Queensland beaches are blessed with a deadly box jellyfish called *Chironex fleckeri*, so it is little wonder that public pressure mounted when the stage 2A dam was filled to capacity from 1989: this fresh waterbody presented a cool haven of safety and tranquillity in which to swim, boat, picnic, fish, watch birds and bushwalk (Figure 9.1).

From 1982 to 1985, some mosquito studies were done at the stage 1 dam (Barker-Hudson *et al.* 1993; Kay *et al.* 1990; Jones *et al.* 1991; Rae 1983). These defined the mosquito species present, the trapping methodologies suitable for their surveillance and established that alphaviruses such as Ross River virus were common. From 1989, the Townsville-Thuringowa Water Supply Board, aided by

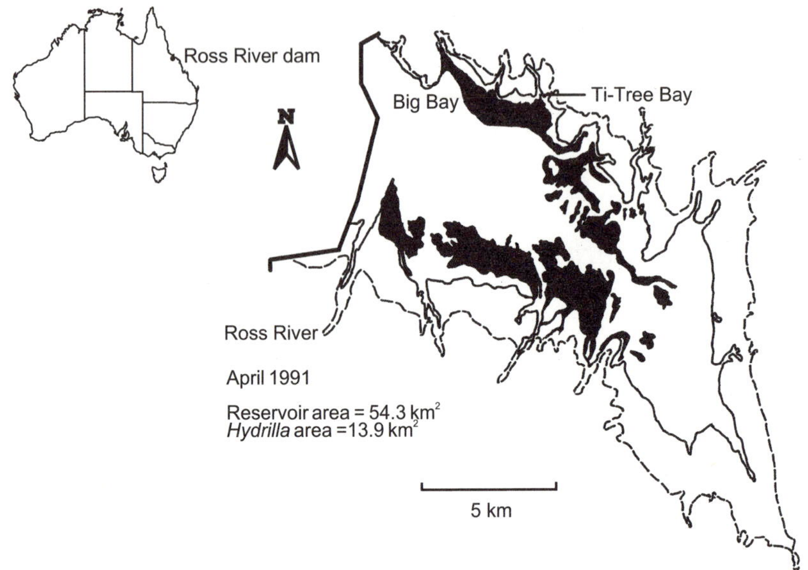

Figure 9.1 The Ross River reservoir showing floating *Hydrilla* beds as black patches (inset eastern coast of Australia).

the federal Land and Water Resources Research and Development Corporation, commissioned a collaborative study between the Queensland Institute of Medical Research and the Zoology Department of James Cook University, Townsville. The objectives of this study of the enlarged waterbody were to define the set of hazards and find management solutions that would facilitate some recreational usage.

9.2 Recreational report

Following a 1978 land use study by the Queensland Department of Local Government, a second report on recreational use of the lake and foreshores was issued (Department of Local Government 1985). The study addressed potential user groups as mentioned earlier and mainly addressed public management, physical disturbance to the foreshore, chemical exhausts and spills and noise. Because of the recognized problem associated with managing night-time activities, it was recommended that most activities should be restricted to daylight hours, except for organized caravanning and camping. Caravans at least were to be parked outside a 400 m buffer zone above full supply level. This of course would probably make caravanning a very seasonal activity as towards the end of the dry season in November, the lake margin may recede as much as 3–4 km. Consequently water frontage may be transposed into expanses of cracked dry mud interspersed with drowned and straggly Chinee apple bush (*Ziziphus mauritania*).

The report recognized the need to minimize disturbance of fauna and flora and suggested that 'swimmer's itch', caused by avian schistosome cercariae, and mosquito-borne viruses should be investigated. Because the 26 km northern boundary, e.g. Big Bay, Antill Creek, had steeper foreshores and deeper water, it was recommended as a primary site for public access. The 7 km western boundary formed by the dam wall was seen as ideal for viewing opportunities of the lake and surrounding hills and mountains, and for water sports. Because of inaccessibility, potential management difficulties and shallowness, the 47 km southern and eastern margins did not offer significant recreational opportunities.

9.3 Tropical itch mite

The stage 1 lake was surrounded with open schlerophyll woodland which afforded kangaroos and wallabies shelter during the hottest times of the day. Part of their exoparasitic fauna is the mite *Eutrombicula macropus*, whose offspring spend part of their life-cycle hanging off grass stems and other vegetative matter waiting to encounter a new host. Much to their misfortune, campers and bushwalkers consequently often find themselves with an itchy rash called 'tropical itch', often around the lines of underclothing.

Prior to the filling of the stage 2 lake, the land in the zone between the stage 1 and stage 2 margins was selectively cleared. This probably diverted the macropods to other wooded habitat. From November 1990 to 1992, 350 litter samples were processed using Berlese funnels and 40 W incandescent bulbs to drive any inhabitants into sample bottles containing 70 per cent alcohol. No *Eutrombicula macropus* were collected. Thus clearing would seem to present an effective management option against this pest, as well as having the other benefits detailed below.

9.4 Mosquitoes and arboviruses

9.4.1 Mosquitoes

From April 1984 to September 1985 (stage 1), the primary questions related to definition of mosquito taxa and the suitability of different methods of catching adult mosquitoes for surveillance purposes. Twenty-six taxa were collected by all night carbon dioxide supplemented light traps or by human bait collections for one hour after sunset (Barker-Hudson *et al.* 1993; Jones *et al.* 1991). The numerically dominant species were *Culex annulirostris* and *Anopheles annulipes* (both species groups), which are traditionally associated with temporary fresh water pools along the lake margins, often among emergent vegetation. Of considerable surprise during September 1985 was the discovery of immatures of these species, plus *Aedeomyia catasticta*, utilizing extensive floating mats of the aquatic weed *Hydrilla verticillata* which sometimes covered 37 per cent of the surface of the lake. This will be discussed later. Two species, *Mansonia uniformis* and *Mansonia septempunctata*, which breed in association with macrophytes such as water hyacinth *Eichhornia crassipes*, became less common from stage 1 to 2. The saltmarsh species *Aedes vigilax* was also collected in reasonable numbers at all localities around the reservoir. This species is known for its wide dispersal powers and was undoubtedly blown in from the extensive intertidal wetlands on the coast.

Thus on the basis of abundance, two taxa – *Culex annulirostris* and *Anopheles annulipes* s.l. – warranted further consideration. The former species is considered to be the major vector of arboviruses in Australia (Russell 1995), transmitting Ross River, Barmah Forest, Kunjin, Kokobera, Alfuy and Edge Hill viruses and Murray Valley encephalitis, as well as dog heartworm. Of these, Ross River is by far the most common arbovirus in coastal northern Queensland, with morbidity approximating 400 cases per 100,000 population. Thus from first principles, this arbovirus and perhaps Barmah Forest, about which little is known, would constitute the greatest hazard to recreational use. Although *Anopheles annulipes* has previously been implicated in malaria transmission at Sellheim during the Second World War, this species group has returned isolated positives of Ross River and Barmah Forest viruses and Murray Valley encephalitis from other parts of Australia. However, no transmission studies have been done on the population from the reservoir. Thus on the evidence to date, it could not be regarded as a major concern at the Ross River dam.

Both *Culex annulirostris* and *Anopheles annulipes* were shown to have seasonal peaks of abundance during the late post-wet season (March to May), with populations building up with the onset of spring (September to October). Spatially, the trapping programme was designed to compare mosquito numbers on the foreshore of the stage 1 lake with two localities expected to be on the margins of the stage 2A lake, with two remote localities (and therefore theoretically unaffected by any water resource project activity) as negative controls. Mosquito numbers (i.e. for those species known to breed at the dam) decreased with distance away from the Ross River dam.

Both light trapping and human bait collections carried out twice per month were reasonable indicators of broad seasonal trends in mosquito abundance. However, the statistical analysis indicated that occasionally the light traps could miss short periods of high biting activity (Jones *et al.* 1991). If greater resolution was required, it was recommended that light traps could be supplemented with animal baited traps, although it is probable that this could be rectified by intensifying the light trapping regimen. Cluster analyses of dam breeding species in both 1984–85 and 1991–93 indicated that light trap catches along the northern (Big Bay, Ti-Tree Bay, Round Island) and western sides (Ross River) gave similar patterns, but the profile towards the east (Antill Creek, Toonpan, Oak Valley) was somewhat different (Barker-Hudson *et al.* 1993; Hearnden and Kay 1995). On this basis, adult mosquito surveillance would therefore need to be based on two localities at either end of the lake.

The lake trapping was continued twice monthly from February 1991, two years after the first filling of the stage 2A reservoir, until June 1993. The trapping locality at Toonpan was essentially the same as for the 1984–85 studies except that for Big Bay was moved a few hundred metres up the incline. Because the expansion from stage 1 to 2A involved extensive clearing of marginal scrub, grassland and forest, almost total control of five mosquito species utilizing tree holes and plant axils (*Aedes alboscutellaris*, *Aedes mallochi*, *Aedes purpureus*, *Aedes quasirubithorax*) or shaded pools (*Uranotaenia nivipes*) occurred. The transformation of temporary wetland with ti-trees (*Melaleuca* spp.), lilies (*Nymphoides indica*, *Nymphaea gigantea*) and submerged plants into an unvegetated muddy foreshore similarly reduced *Mansonia* spp. and *Coquillettidia crassipes*, whose larvae depend on attachment to arenchymatous or lacunate macrophytes. Larvae of these genera have pointed reinforced tips to their siphons which are used to pierce these plants to breathe.

Because of the devastating nature of the inundation and the time required for new breeding habitat to re-establish, mosquito populations increased through to the end of 1993 but the mean abundance of adult *Culex annulirostris* had not changed significantly from stage 1 levels. The trend for this species and for *Anopheles annulipes* was upward, and one can only speculate on population levels when the marginal vegetation has fully established. Due to the extensive loss of marginal vegetation and the creation of expanses of shallow muddy pools, especially towards Toonpan, *Anopheles amictus* and *Aedes normanensis* populations increased by 36-fold and 282-fold, respectively (Figure 9.2). The ramifications of this are interesting as *Aedes normanensis* is well recognized as a vector of Ross River virus and Murray Valley encephalitis, especially inland where *Anopheles amictus* (probably another species complex) has been the source of Ross River, Barmah Forest and Edge Hill viruses.

Control of mosquitoes is usually directed at removal of breeding habitat (source reduction) or aimed at larvae which often aggregate in large numbers in discrete sites. *Aedes normanensis* is ephemeral and its desiccation-resistant eggs characteristically hatch in response to wet season rainfall filling up temporary pools. Plague numbers appear one month and may be gone the next. More accurate definition of these breeding sites, particularly at Toonpan, Antill Creek and Ross River, is required before control options can be considered.

As already mentioned, the clearing process created vast expanses of bare muddy pools, particularly at the north-eastern end (e.g. Toonpan). As the lake gradually receded during the dry season, ideal breeding sites were created and populations increased through spring (from September) and also in the late wet season (March to April) when dry sites were refilled by rainfall. Thus, although the land clearing had benefits in eliminating tropical itch mites and some minor mosquito species, it probably paved the way for population growth of *Aedes normanensis* and *Anopheles amictus*. This could possibly be considered a dubious swap, although time will tell. Little is known of their biology and their flight range, the latter being of obvious importance to recreational activity at the other end of the lake. Fortunately, however, they are mainly active at night.

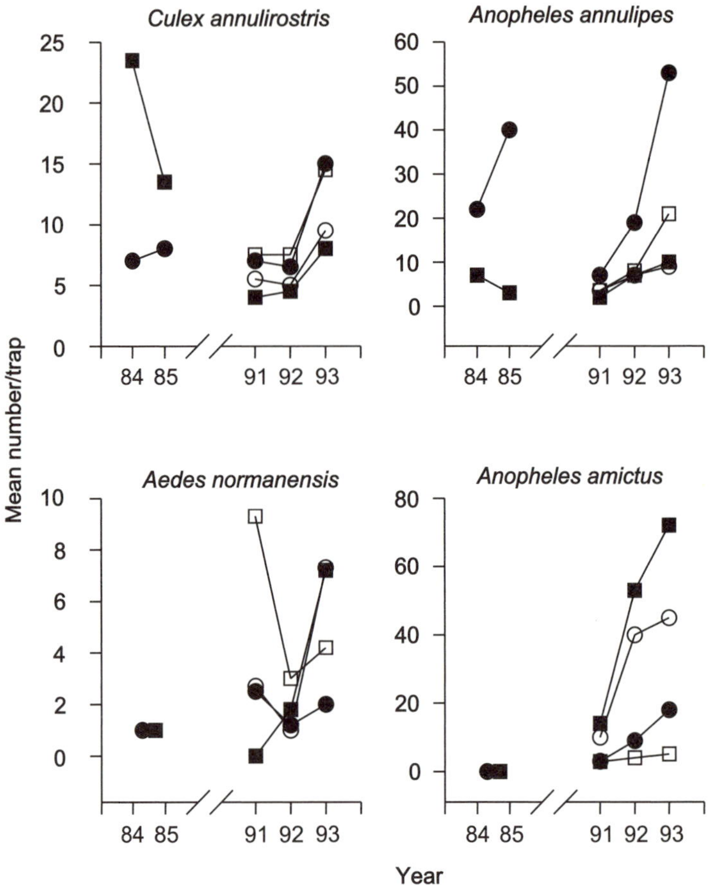

Figure 9.2 Comparison of overall means of adult female mosquitoes collected by light trap, 1984–1985 and 1991–1993 (January-June). ■ Ross River; ● Antill Creek; □ Toonpan; ○ Big Bay.

The breeding habitat within the boundaries of the reservoir can be divided into three: the *Hydrilla* beds; the lake margins; and further away from the lake, ponded creeks, pools and hoof prints. Although mosquito larvae were collected in muddy pools and along the foreshore, the presence of *Hydrilla* mats floating on the surface constituted reason for concern. During January to March and September 1985, average mosquito densities in this weed at Ti-Tree Bay and Round Island were 272 immatures/m^2, 90 per cent being *Culex annulirostris* (Kay *et al.* 1990). On the basis of area, this equated to in excess of 1 billion juveniles from *Hydrilla*.

From April 1991, similar assessments of the productivity of floating *Hydrilla* beds adjacent to the northern and southern shores were made (Figure 9.1). However, because the lake was larger, aerial photography was used to estimate the extent of the floating algal mats. This was then scanned into a computer, digitized and estimates made. During April and November 1991 and in September 1992, *Hydrilla* covered between 13.3 and 16.6 km^2 of the stage 2A lake, and this was estimated to contain populations of 5.6 billion, 275 million and 513 million immatures, respectively, mainly *Culex annulirostris* and *Anopheles annulipes* as before. However, whereas *Culex annulirostris* comprised 90–98 per cent of aquatic stages during 1985–86, now *Anopheles annulipes s.l.* comprised 43.7 per cent of all the immatures identified. Although natural mortality will reduce the numbers actually reaching adulthood, these numbers are so high that some form of control is indicated.

As discussed previously, one can only speculate on the abundance of mosquito larvae that will utilize marginal emergent vegetation when it develops fully along the foreshores of the stage 2A lake. At present, from the 1991–93 data, the average number of mosquito immatures based on transects 5 m wide was 85.7/m^2. Because the shoreline contains bare patches unsuitable as mosquito habitat, even if a 5 m wide swathe the entire length of the 115 km perimeter was producing, the theoretical maximum is only 49 million. This constitutes only 2.3 per cent of the average number of 2,149 million estimated for *Hydrilla* over three surveys. From transects done in February 1986, the average was 528/m^2 with a maximum of 2,704 larvae, mainly *Culex annulirostris* /m^2. This would suggest that with the gradual stabilization of the Ross River dam (stage 2A), *Culex annulirostris* will increase and, along with it, the presence of medically-important arboviruses.

Given the relative importance of *Hydrilla* as mosquito breeding habitat at present, we can ask two questions:

- What is the expected impact of controlling mosquito larvae in the *Hydrilla*?
- How should it be done?

The first question was approached by statistical correlation of the varying absolute abundances of immatures in the *Hydrilla* from 1991–93 with the catches of adults from the light trap surveillance programme. Although correlations approached statistical significance in some cases, generally the results for the prime recreation site, Big Bay, indicated that the adult mosquitoes were coming from various habitats and probably also from breeding sites outside the boundaries of the reservoir.

Mosquitoes such as *Anopheles annulipes* are known to disperse 5–16 km (Fenner and Ratcliffe 1965), whereas *Culex annulirostris* has been recorded 12 km away from release points (Bryan *et al.* 1992). *Aedes vigilax*, the saltmarsh vector of Ross River and Barmah Forest viruses, can easily travel 50 km and is consequently distributed more or less evenly over the reservoir area (Barker-Hudson *et al.* 1993).

Given that nothing is perfect in this world, we still consider that control of mosquitoes in *Hydrilla* is worthy of consideration. Should we simply get rid of the *Hydrilla*, which acts as a water purifier and a food source and haven for the abundant wildlife at the reservoir? We think not. This option will be discussed later.

9.4.2 Arboviruses

From 1983 to 1987, domestic chicken flocks were maintained as sentinels (Figure 9.3) and bled for the presence of antibodies to both alphaviruses (e.g. Ross River virus) and flaviviruses (e.g. Murray Valley encephalitis, Kunjin viruses). These data showed that alphavirus activity occurred annually but this infection was somewhat unpredictable and did not necessarily correlate with the zones of highest mosquito abundance closest to the reservoir. Flavivirus infection occurred far less frequently. From 1990–93, 51,497 adult female mosquitoes were processed for the presence of alphaviruses using enzyme immunosorbent assay and polymerase chain reaction methods (Kay *et al*. 1996; Oliveira *et al*. 1995). Ten isolates of Ross River, one of Barmah Forest and two of Sindbis virus strains were recovered from *Aedes normanensis*, *Anopheles amictus* and *Culex annulirostris*, which indicated that they were probably local vectors. All virus stains were recovered from mosquitoes collected during the wet seasons (February or March) of 1991 or, especially, 1992.

Relative hazard was assessed for four Ross River dam localities: Big Bay (north); Antill Creek (north-east); Toonpan (east); and Ross River (south). These were compared with more limited data for the greater Townsville area (Kay *et al*.

Figure 9.3 Sentinel chicken flocks were maintained around the reservoir margin to monitor seasonal antibody conversion to arboviruses.

1996). Based on the prevalence of key vector species and their abundance and that of the viruses recovered, it was concluded that Big Bay, originally recommended as a prime site for recreational development by the Department of Local Government in 1985, actually presented lower risk than any other locality. Antill Creek also proved relatively safe in terms of mosquito-borne infections, whereas Toonpan during the wet season was a place to be avoided. Both Ross River and the environs of Townsville offered intermediate risk, the latter due to large numbers of saltmarsh mosquitoes breeding in intertidal wetlands.

9.5 Snails and swimmer's itch

Schistosome dermatitis, known as swimmer's itch, is a common global problem for users of recreational swimming areas in water resource developments. The rash is caused by free living larvae called cercariae (Figure 9.4) of parasitic flukes which burrow into exposed parts of the body. Normally the life-cycle involves water birds such as ducks and pulmonate snails, so infection of humans is accidental. A large number of cercariae may penetrate the skin where they die but cause a localized allergic reaction in sensitized persons.

In northern Australia, swimmer's itch (*Trichobilharzia*) has been traditionally associated with *Austropeplea* (= *Lymnaea*) *lessoni* (= *vinosa*) although two planorbid snails, *Amerianna carinata* and *Gyraulus stabilis*, have also been identified as intermediate hosts in Lake Moondarra near Mt Isa, Queensland. Our recent data implicates *Gyraulus gilberti* at the Ross River dam. Snails are also commonly infected with other trematode cercariae, mainly echinostomes, strigeids/diplostomids and clinostomids.

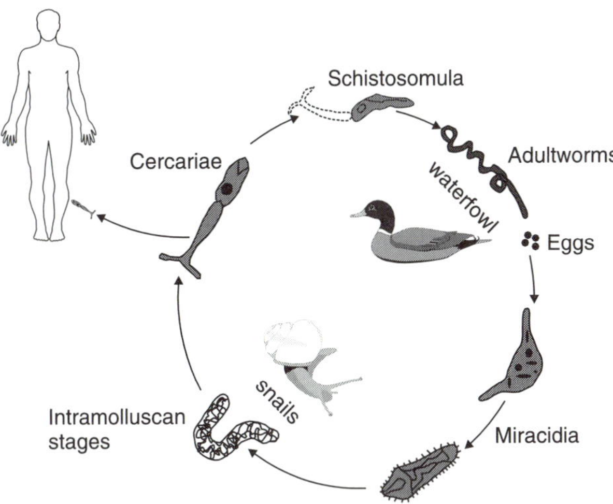

Figure 9.4 The cercaria of *Austrobilharzia* penetrate through skin and cause 'swimmer's itch'. The parasitic cycle involves waterfowl and snails as intermediate hosts.

Because of previous isolated reports of swimmer's itch at the stage 1 lake, periodic surveys by sweep netting commenced in July 1990. For each of six localities covering three distinct habitat types (open bays within the lake, along the margins of permanent creeks, temporary ponds), 1 m^2 quadrats containing all vegetation types were scoured for snails. *Austropeplea* snails were not present in the lake proper until February 1991, but in November 1990 they were first located in ponds along the north-east, east and south-western shorelines. Egg masses were often found attached to the undersides of nardoo (*Marsilea mutica*) and sometimes wrapped around the stalks and ventral surfaces of the water lily, *Nymphaea gigantea*. Thus its absence from the lake was attributed to the lack of established vegetation in the stage 2A lake, and from this we developed a working hypothesis that host snails were possibly vegetation-specific. Thus to facilitate recreational use, control of infected *Austropeplea* could be achieved by simply clearing the appropriate water plant.

By July to August 1991, however, schistosome-infected *Austropeplea* were collected from various types of vegetation along the margins of Ross River, close to the lake. A few *Amerianna* and *Gyraulus gilberti* were found in Ti-Tree Bay contiguous with Big Bay and Round Island, which were still negative for snails. By February 1992, planorbids were present in all three habitat types, with *Austropeplea* in two, i.e. ponds and creeks around the lake.

Until 1993, 2,365 snails were dissected to detect both patent and pre-patent *Trichobilharzia* infection. Four different species of snails were identified, size classed according to shell length or width using vernier calipers. Snails were crushed on a microscope slide or in a Petri dish with a few drops of water under a warm light. A heavy infection of cercariae is evident to the naked eye, but any worm-like animals were removed on to another slide, stained with two to three drops of 0.1 per cent neutral red dye, covered with a coverslip and examined microscopically. Schistosome cercariae are distinctive with their eye spots, forked tail and presence of oral and ventral suckers (see Figure 9.4). Schistosomes were recovered from 4.5 per cent and 1.7 per cent of Austropeplea and *Gyraulus gilberti* snails, but not from *Amerianna* nor *Thiara*.

In terms of management solutions, several questions seemed paramount:

- Which habitat types presented the greatest (and lowest) risk?
- Which time of the year presented the greatest hazard?
- Can certain indicators be used to predict infection?

Statistical analysis of the presence and abundance of *Austropeplea* snails did not correlate with any particular vegetation type (Hurley *et al.* 1995) but was connected with vegetation generally. There was no clear-cut relationship with snail density and physicochemical parameters including temperature, biomass of periphyton or with percentage surface coverage. However, highest densities of *Austropeplea lessoni* (45/m^2) and *Amerianna carinata* (190/m^2) were recorded in temporary ponds where the periphyton biomass was greatest. Periphyton is mostly composed of diatoms, epiphytic algae, bacteria, fungi and protozoa and builds up as a food source for snails on submerged waterplants and on the undersides of floating plants. Highest densities of *Gyraulus gilberti* (15/m^2) occurred in permanent creeks such as Ross River. Thus, rather fortuitously, likely recreational sites such as Big Bay and Ti-Tree Bay along the northern foreshore were relatively safe, especially if some vegetation clearing was added. *Austropeplea* was never collected from bare areas.

Although infected snails were collected at all times of the year, the greatest densities of *Austropeplea* occurred during the late dry season. Blair and Finlayson (1981) noted the catastrophic effect of heavy wet season rainfall on *Austropeplea* populations at the Ross River dam in 1978 due to substrate scouring and vegetation clearing. It is likely that since *Austropeplea* were mainly associated with ponded habitats, maximum densities were influenced considerably by dry season habitat shrinkage as well as by food concentration. In contrast, peak densities of *Amerianna* and *Gyraulus* were recorded during the late wet seasons of 1992 and 1993.

In terms of surveillance, the prevalence of infection within each size class of snail provided a handy indicator to streamline efforts. For *Austropeplea lessoni*, for example, 80 per cent of snails of 20 mm or greater shell length were infected with trematodes, 39 per cent of these being with *Trichobilharzia* spp. For the Ross River dam, smaller snails, i.e. < 3 mm for *Gyraulus gilberti*, < 6 mm for *Amerianna carinata* and < 10 mm for *Austropeplea lessoni*, need not be collected (Hurley *et al.* 1994).

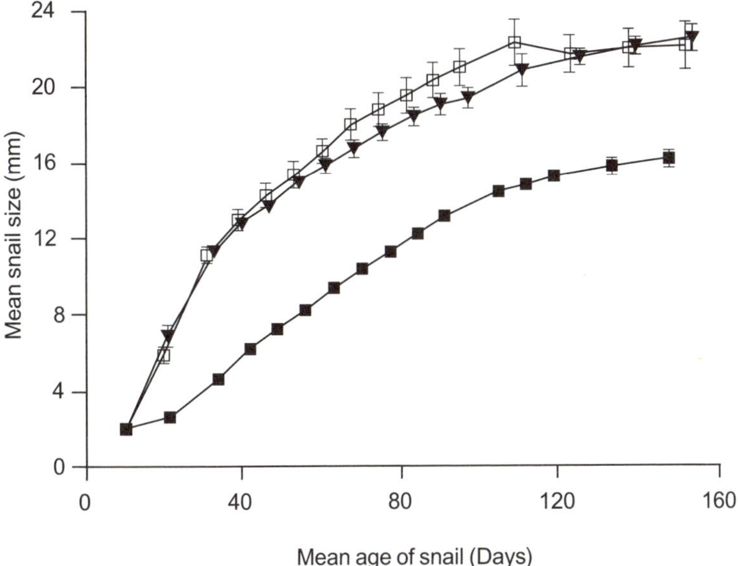

Figure 9.5 Growth of *Austropeplea lessoni* snails in relation to different water temperatures (15°C ■; 25°C □; 30°C ▼).

As an adjunct to this, egg masses of *Austropeplea* were hatched out and reared in constant temperature rooms at 15°C, 25°C and 30°C with weekly changes of water and vegetation (Figure 9.5). Shell length was measured weekly until time of reproduction. At 15°C the snails grew slower but lived longer, but at 25°C and 30°C, there was little difference in growth rates, although those at 25°C were marginally larger at equivalent periods. Although water temperatures at the Ross River dam do occasionally drop to 16°C on occasions, generally they average 25–28°C (Hurley *et al.* 1995). Thus from this, an *Austropeplea* of 12 mm shell length collected during summer will be around one month old and capable of reproducing. One of 20 mm at either 25°C or 30°C water temperature would be approximately 100 days old. On this basis, it is suggested that monitoring could be comfortably done every two to three months.

9.6 Management options

9.6.1 General conclusions

There are several other lakes, man-made or otherwise in northern Queensland, that support diverse recreational activities without apparent mishap. All are subjected to tropical conditions conducive to year round production of mosquitoes, snails, mites and pathogens. What is different about the Ross River dam stage 2A is its shallowness and proximity to large human populations.

Nevertheless, the studies carried out in two blocks (1983–1987 and 1990–1995) have defined its mosquito and alphavirus hazard as considerable but no greater in the northern and north-eastern areas of Big Bay, Ti-Tree Bay, Round Island and Antill Creek than that experienced by local residents in everyday life. The relative hazard would change considerably, however, if the responsible local authorities ever decided to mount a broadscale aerial control programme against larval *Aedes vigilax*, which breed in the extensive intertidal wetlands.

Restriction of activities to daylight hours will not only facilitate easier control of the public but will also reduce exposure to key vector species such as *Culex annulirostris*, *Anopheles amictus* and *Aedes normanensis*. However, who takes the responsibility for an estimated 5 billion mosquito larvae found periodically in the floating *Hydrilla* beds? As discussed, both *Culex annulirostris* and *Anopheles annulipes* are quite capable of dispersing from the reservoir into the urban populace. Recreational management issues are probably far less complicated than the moral issues.

Whereas land clearance prior to the flooding of the stage 2A lake was effective in controlling tropical itch mites and some mosquito species, it also probably effected a redistribution of the kangaroos and wallabies, known to be most effective intermediate hosts of some arboviruses, including Ross River and the often fatal Murray Valley encephalitis. They have probably been driven towards the quieter eastern areas around Toonpan, where in 1992 Ross River virus was detected in wet season *Aedes normanensis* at rates as low as 1:217 mosquitoes.

What is the impact of such ecological change and what will it look like in the future?

9.6.2 Mosquito or aquatic plant control?

The options for control of aquatic plants such as *Hydrilla* are mechanical, biological, chemical, or a combination of these methods. The objective of aquatic weed control should be to control growth sufficiently to permit the water to be used in the desired way but without a change in the balance of species (Bill 1977).

Aquatic plants are only weeds if they pose a major nuisance or hazard. Clearly there is a case as mentioned previously for clearing buffer zones to mitigate against swimmer's itch or to facilitate boating and safe swimming.

Aquatic plant growth generally relies upon nutrient availability, light availability, adequate physicochemical characteristics and habitat stability. Nutrient availability relies upon substrate type and the presence of dissolved organic and inorganic matter. Light intensity decreases with depth to the point where the energy acquired by photosynthesis cannot meet the energy requirement of vegetation and plant growth ceases. The interrelationships of key factors such as depth, wave exposure, littoral slope and sediment characteristics are complex (Duarte and Kalff 1990), although slopes of greater than 15° are regarded as the first limit to plant growth and the second is depth. The Ross River reservior is shallow with an average depth of less than 3 m, which explains why *Hydrilla* beds sometimes cover up to 37 per cent of the surface area of the lake.

Bill (1977) discussed a protocol for deciding the best and most effective control measures to be used and outlined a checklist of questions.

- To what extent is plant growth responsible for the particular problem, e.g. reduction of channel capacity, interference with recreational use?
- Are chemical methods of control more suitable than mechanical or biological methods, or could more than one method be used?
- What is the most economical long-term approach?
- What degree of control is required to provide adequate relief from the particular problem?
- If chemical methods are most appropriate, which material is likely to be most effective and how should it be used? Are residues of chemicals in the water following a treatment likely to be detrimental to human health or to fish, wildlife or irrigated crops?
- Is it desirable to retain some plants for the benefits of fish and waterbirds?

Biological control is not the universal solution to all pest problems, but it may be applied to a vast array of problems and when effective it is the most satisfactory and economical form of control. The state of Queensland has generous expertise in this area, with the CSIRO Division of Entomology – Lands Department group in Brisbane boasting spectacular success against *Salvinia* and *Eichhornia*, and near the reservoir at James Cook University a USDA unit was involved in successes with the Tennessee Valley Authority (TVA) (see Chapter 12) using a range of stem-boring and leaf-mining insects (Balciunas *et al.* 1993).

One might consider the herbivorous grass carp *Ctenopharyngodon idella*, originally from China, more as a harvester than a biological control agent. This fish grazes on submerged weeds such as *Hydrilla*, *Myriophyllum*, *Chara*, *Potamogeton* and *Ceratophyllum*, and at stocking rates of 75 fish/ha control is rapidly achieved. Some introductions in the USA have resulted in removal of all vegetation (Leslie *et al.* 1987), and in the Australian context the use of sterile (triploid) fish (Cassani and Canton 1985) could be the only consideration. However, in view of the damage already done by grass carp to some inland waterways in Australia, it is suspected that this option would be greeted with horror.

Mechanical control involves the physical removal of weeds from a problem area and is useful in situations where the use of herbicides is not practical or poses risks to human health or the environment. Mobile harvesters sever, lift and carry plants to the shore. Most are intended for harvesting submerged plants, though some have been designed or adapted to harvest floating plants. Handling the harvested weed is a problem because of their enormous water content, therefore choppers are often incorporated into harvesting machinery design. However, many mechanical harvesters have a small capacity and the process of disposing of harvested plant material is time-consuming. Any material that remains may affect water quality during the decay process by depleting the water of oxygen. Furthermore, nutrients released by decay may cause algal blooms (Mitchell 1978). Another disadvantage of mechanical removal is that disturbance often promotes rapid new growth and germination of seed, and encourages the spread of weed by fragmentation.

Some direct uses of macrophytes include the following: livestock food; protein extraction; manufacture of yeast; production of alcohol and other by-products; the formation of composts, mulches and fertilizers; and use for methane generation (Williams 1977).

Herbicides either kill on contact, or after translocation through the plant. Some are residual and retain their toxicity for a period of time. Where herbicides are used for control of plants, some contamination of the water is inevitable (Bill 1977). The degree of contamination depends on the toxicity of the material, its fate and persistence in the water, the concentration used and the main purpose served by the water. After chemical defoliation of aquatic vegetation, the masses of decaying organic debris produced can interfere with fish production. Several factors must be taken into account when selecting and adapting herbicides for aquatic purposes, including: type of water use; toxicity of the herbicide to humans, fish, stock, and wildlife; rate of disappearance of residues, species affected and duration of control; concentration of herbicide; and cost (Bill 1977).

The TVA has successfully used EPA-approved herbicides such as Endothall, Diquat, Fluridone and Komeen against *Hydrilla* (Burns *et al.* 1992), and a list of

(a)

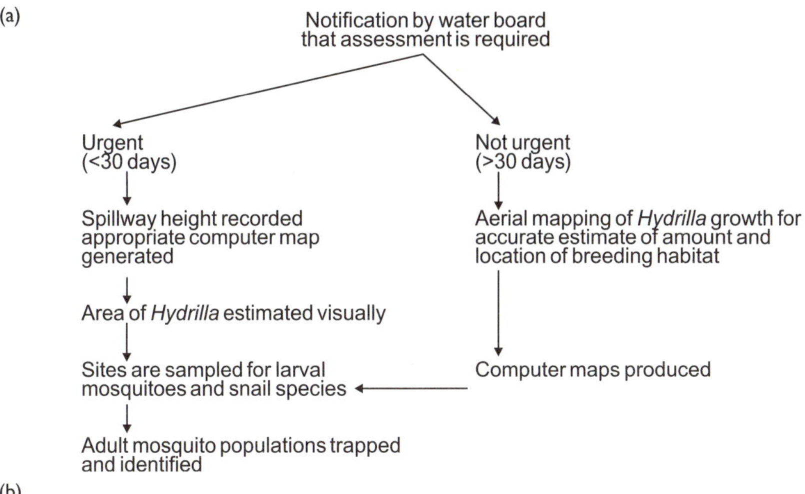

(b)

Risk assessment - Ross River Dam Date: 27 November 1994 Spillway height: 29.56m

Potential breeding area			Mosquitoes (Numbers per trap)			Snails		
Shoreline (km)		40.59	Species	Big Bay	Toonpan	Number/m^2		Nil
Hydrilla (km^2)		0.00	*Culex annulirostris*	3.0	0	Low	Medium	High
Low	Medium	High	*Anopheles annulipes*	0.5	0	Percent infection		NA
			Anopheles amictus	0.88	0			
			Aedes normanensis	0	0	Low	Medium	High

Risk assessment for 30 day period from date of issue:	
Arbovirus transmission	Swimmer's itch
Minimal: Major habitat for breeding (floating *Hydrilla verticillata*) absent. Catches of adult females very low for the major vector species	Minimal to absent: Aquatic vegetation around facilities at boat ramp very sparse. Current habitat does not support populations of host snail species.

Figure 9.6 (a) Flow chart for assessing risk of mosquito- and snail-associated problems to recreational users at Ross River dam. (b) Risk assessment sheet supplied by technical group to the Townsville-Thuringowa Water Supply Board.

approved products for Australia is available from the Land and Water Resources Research and Development Corporation. However, the option we favour fits in with TVA policy on aquatic weeds that do not conflict with public usage, i.e. leave the weed and treat the pests within it. Therefore we would recommend that during the period of greatest risk for arboviruses and when mosquitoes are at peak abundance (March to May following the wet season), biorationals such as *Bacillus thuringiensis israelensis* or s-methoprene should be applied. At the dosages used to kill mosquitoes, non-target organisms are safe. Both of these products, especially *Bacillus*, break down quickly and should therefore present no hazard to water quality.

Given that a site such as Big Bay may become a mecca for those wishing to swim, sail or even fish, a surveillance programme and some environmental modifications are necessary. The deeper open waters of this bay coupled with a vegetation-free foreshore as a buffer zone, perhaps 400 m on either side of a swimming zone, should minimize or even negate swimmer's itch. Adjacent *Hydrilla* and other macrophytes will require clearing as these will also present a physical hazard to swimmers and watercraft.

The monitoring programmes could ideally be done three to four weeks prior to extensive public usage to allow time for any remedial action. The prevalence of key mosquito species and of large *Austropeplea* (and *Gyraulus* and *Amerianna*) snails can be established quickly as can cercarial infection in the snails. If it is found necessary to establish infection rates in mosquitoes, the newly developed Ross River and Barmah Forest virus testing procedures using mosquito cell cultures and enzyme immunosorbent assay (Oliveira *et al.* 1995) would require six days processing time. This offers considerable economy over previous methods using intracerebral inoculation of baby suckling mice.

We would suggest that prior to selected recreational events, especially those from March to May, the Water Supply Board should initiate the action shown in Figure 9.6. The information supplied in response to a request should be communicated to recreational users to ensure that they are aware of the risks. Perhaps mosquito, arbovirus, and schistosome status could be displayed in the same way as fire hazard status is commonly indicated.

It would be remiss of us to generate the impression that we had all the answers to the Ross River dam. The stage 2A lake and its surrounds are undergoing a process of ecological change and realization of this must remain paramount. There are issues relating to mosquito biology and behaviour and to do with snail ecology generally that would repay further study. Thus further selective monitoring and research should not be forsaken.

References

Balciunas, J.K., Burrows, D.W. and Purcell, M.F. (1993) Final report on the Australian surveys (1985–1992) for insect biological control agents of *Hydrilla verticillata*. U.S. Army Corps of Engineers, Waterways Experiment Station, Vicksburg, USA.

Barker-Hudson, P., Kay, B.H., Jones, R.E., Fanning, I.D. and Smythe, L.D. (1993) Surveillance of mosquitoes and arbovirus infection in the Ross River dam (stage 1), Australia. *Journal of the American Mosquito Control Association*, **9**, 389–399.

Bill, S.M. (1977) Learning to live with aquatic plants. In: *The Menace of Water Hyacinth and other aquatic weeds*. Water Research Foundation of Australia, Adelaide.

Blair, D. and Finlayson, C.M. (1981) Observations on the habitat and biology of a lymnaeid snail *Austropeplea vinosa* (Gastropoda: Pulmonata), and intermediate host for avian schistosomes in tropical Australia. *Australian Journal of Marine and Freshwater Research*, **32**, 757–767.

Bryan, J.H., O'Donnell, M.S., Berry, G. and Carven, T. (1992) Dispersal of adult female *Culex annulirostris* in Griffith, New South Wales, Australia. A further study. *Journal of the American Mosquito Control Association*, **8**, 398–404.

Burns, E..R., Bates, A.L. and Webb, D.H. (1992) *Aquatic Plant Management Program. Current Status and Seasonal Workplan 1992*. Tennessee Valley Authority, Muscle Shoals, USA, pp. 66.

Cassani J.R. and Canton, W.E. (1985) Induced triploidy in grass carp, *Ctenopharyngodon idella* Val. *Aquaculture*, **46**, 37–44.

Department of Local Government (1985) *Ross River Dam Catchment Area, Stage 2 Development. Report on Recreational Use of Lake and Foreshores; Buffer Zone* Government Printer, Brisbane.

Duarte, C.M. and Kalff, J. (1990) Patterns in the submerged macrophyte biomass of lakes and the importance of the scale of the analysis in the interpretation. *Canadian Journal of Fish Aquatic Science*, **47**, 357–363.

Fenner, F. and Ratcliffe, F.N. (1965) *Myxomatosis*. Cambridge University Press, Cambridge.

Hearnden, M.N. and Kay, B.H. (1995) Changes in mosquito populations with expansion of the Ross River reservoir, Australia from Stage 1 to Stage 2a. *Journal of the American Mosquito Control Association*, **11**, 211–224.

Hurley, M., Hearnden, M., Blair, D. and Kay, B.H. (1994) Larval trematodes in fresh-water snails at Ross River Dam, northern Australia, with emphasis on *Trichobilharzia* spp., causative agents of swimmer's itch. *Australian Journal of Marine and Freshwater Research*, **45**, 563–567.

Hurley, M., Hearnden, M.N. and Kay, B.H. (1995) The distribution of aquatic pulmonate snail species in contrasting habitats within the Ross River reservoir (Stage 2A), tropical northern Australia. *Australian Journal of Marine and Freshwater Research*, **46**, 1033–1038.

Jones, R.E., Barker-Hudson, P. and Kay, B.H. (1991) Comparison of dry ice baited light traps with human bait collection for surveillance of mosquitoes in northern Queensland, Australia. *Journal of the American Mosquito Control Association*, **7**, 387–394.

Kay, B.H., Barker-Hudson, P., Piper, R.G. and Stallman, N.D. (1990) Arbovirus disease and surveillance methodology related to water resource development in Australia. *International Journal of Water Resources Development*, **6**, 95–103.

Kay, B.H., Hearnden, M.N., Oliveira, N.M.M., Sellner, L. and Hall, R.A. (1996) Alpha-virus infection in mosquitoes at the Ross River dam, north Queensland, 1990–1993. *Journal of American Mosquito Control Association*, **12**, 421–428.

Leslie, A.J., Van Dyke, J.M., Hestand, R.S. and Thompson, B.Z. (1987) Management of aquatic plants in multiuse lakes with grass carp (*Ctenopharyngodon idella*). *Lake and Reservoir Management*, **3**, 266–276.

Mitchell, D.S. (1978) Aquatic Weeds. In: *Australian Inland Waters*. Australian Government Publishing Service, Canberra.

Oliveira, N.M.M., Broom, A.K., Lindsay, M.D.A., Mackenzie, J.S., Kay, B.H. and Hall, R.A. (1995) Enzyme immunoassays for the rapid detection of Ross River virus in infected mosquitoes. *Clinical and Diagnostic Virology*, **4**, 195–205.

Rae, D. (1983) The mosquito larvae of Ross Dam with particular reference to the ecology of *Culex annulirostris*. BSc (Honours) thesis. Zoology Department, James Cook University of Northern Queensland, Townsville.

Russell, R.C. (1995) Arboviruses and their vectors in Australia: an update on the ecology and epidemiology of some mosquito-borne arboviruses. *Review of Medical and Veterinary Entomology*, **83**, 141–158.

Williams, W.D. (1977) The Australian inland aquatic environment: A background review for management purposes. In: *The Menace of Water Hyacinth and other Aquatic Weeds. Water Research Foundation of Australia*, Adelaide.

Chapter 10

Artificial wetlands and mosquito control in Australia

Richard C. Russell

10.1 The general scenario

Wetlands, whether natural or artificial, are water resources and as such attract a variety of organisms both vertebrate and invertebrate that may cause disease. Natural wetlands have a biogeochemical cycling function, trapping a variety of substrates for long periods and purifying the water that passes through them (Breen 1989; Hammer 1992). Artificial wetlands are now being constructed in many localities to 'polish' urban wastewaters by removing organic and inorganic contaminants before direction of the waters into natural drainage systems. In urban areas, wetlands offer the additional advantage of being seen as attractive 'greening' features, providing passive recreation areas and desirable assets such as wildlife refuges.

'Artificial' or 'constructed', rather than natural wetlands, are encouraged for water pollution control in south-eastern Australia because natural wetlands can be adversely affected and become degraded, contributing to other problems in water catchments (Dunkerley and Carter 1992). An important objective for the development of constructed wetlands in south-eastern Australia is the reduction of nutrients entering waterways in order to improve river water quality and reduce the likelihood of algal blooms (Mitchell *et al.* 1994). For instance, in New South Wales (NSW), concern for nutrient pollution leading to both toxic and non-toxic blue–green algal blooms has led the Department of Land and Water Conservation to develop an algal management strategy that includes construction of a range of artificial wetlands for sewage, storm water and agricultural run-off treatment in many parts of the state (Verhoeven and White 1994).

Of course, technologies for constructed wetlands have been under development for more than twenty years now, with consideration of the following:

- stocking of lagoons with floating plants, e.g. water hyacinth (*Eichhornia crassipes*), which are harvested for animal fodder or biogas generation
- flooding fields or constructed ponds and introducing or using pre-existing emergent vegetation
- subsurface flows through soil, sand, or gravel planted plots which, unlike the first two systems, might produce less odour and fewer mosquitoes.

Free water surface flow wetland systems were popular early on in north America; subsurface flow systems were developed in Europe where there is concern about lower temperatures and availability of land. Both have been constructed in Australia, and research has concentrated on the efficiency of the systems containing emergent vegetation. Various plants can be used in constructed wetlands; the common reed (*Phragmites*) is used in artificial wetlands worldwide because of its great volume of root/rhizome zone per surface area of reed bed, but cumbungi (*Typha*) and the rushes *Scirpus* and *Schoenoplectus* are used in many countries. In Australia, various species have been found to be suitable for processing wastewaters (Finlayson and Chick 1983); in a recent report on a subsurface flow wetland receiving secondary treated sewage effluent at Wodonga, Victoria, there was significant improvement in effluent quality in plots with river clubrush (*Schoenoplectus validus*) and giant rush (*Juncus ingens*) (Thomas *et al.* 1994).

The 'industry' is in a 'growth phase'. Currently, constructed wetlands research at the Water Research Laboratory, University of Western Sydney – Hawkesbury, contains more than twenty subprojects with the Cooperative Research Centre for Waste Management and Pollution Control Limited (Roser and Bavor 1994). The latter venture involves three universities, three state departments including the Environmental Protection Authority, the Commonwealth Scientific and Industrial Research Organization (CSIRO) Division of Water Resources and four commercial companies, and is mainly directed at development and application of environmental technologies, especially with reference to the commercial world. This intersectoral approach is commendable, because in the past some designers have been firmly blinkered to consider only water quality issues.

10.2 Wetlands and mosquitoes

While the basic knowledge to design and operate wetlands has been developed, a major drawback has arisen in some areas because of problems with mosquitoes colonizing the systems. This use of wastewater provides ideal mosquito habitat: shallow water with various types of submerged, emergent and floating vegetation.

Most of the literature and comment on artificial wetlands and mosquitoes has emanated from the USA, with its longer and more extensive experience with mosquito control, environmental concerns and wetlands management, but there have been few detailed field studies in constructed wetlands. Constructed wetlands duplicate natural wetlands and should be seen to present similar, if not greater, opportunities for mosquito production through their relative lack of clean shorelines and open/deep waters, which are less productive of mosquitoes because of predator and wave action.

As simple models for constructed wetlands, rice fields have been shown to be capable of generating large populations of important pest and disease vector mosquitoes in various parts of the world (Lacey and Lacey 1990), and have provided data that confirm the importance of managing the mosquito populations.

Even more representative of constructed wetlands are wild rice fields, where the levels of organic matter are greater and the vegetation is more representative of many constructed wetlands (Kramer and Garcia 1989). If wastewater is added to the equation, the productivity can increase substantially; Schaefer *et al.* (1983) found that effluent discharging into rice fields in a number of localities in California resulted in the production of high populations (e.g. thirty times that of a comparable but non-effluent irrigated field) of *Culex tarsalis*, a major pest and vector of several important arboviruses, and they noted that unless the practice of providing such habitats was eliminated, 'a public health catastrophe' may eventuate.

The species of greatest concern in North American constructed wetlands are arguably *Culex pipiens*, *Culex quinquefasciatus* and *Culex tarsalis*, recognized as vectors, variously, of the arboviruses St Louis encephalitis, eastern equine encephalitis and western equine encephalitis. However, other species of significance as pests or associated with disease pathogens (e.g. the arboviruses mentioned above, the California group encephalitides, dog heartworm, fowl pox and bird malaria), such as *Culex erythrothorax*, *Culex quinquefasciatus*, *Culex restuans*, *Culex salinarius*, *Culex stigmatosoma*, *Culex territans*, *Culiseta inornata*, *Coquillettidia perturbans*, *Aedes vexans*, *Aedes squamiger*, have also been recorded in these habitats (Mortenson 1983; Schaefer *et al.* 1983; Tennessen and Painter 1990). Indeed, there are reports from California that mosquito production has been a significant problem in constructed wetlands and is jeopardizing the development of the technology; a number of pilot plants no longer operate because of mosquito problems, and the issue must be resolved before constructed wetlands can be accepted by health officials and the general public (Dinges 1978; Ebipane *et al.* 1993; Eldridge and Martin 1987; Martin and Eldridge 1989; Mortenson 1983).

As another example, elsewhere in the USA, the Tennessee Valley Authority, set up in 1933 (see Chapters 11 and 12), has given high priority to control of mosquitoes in constructed sewage wetlands. At three small communities in western Kentucky (Tennessen and Painter 1990), sewage treatment facilities were failing to meet permit limitation for discharging biological oxygen deficit (BOD), total suspended solids, ammonia-nitrogen and faecal coliforms. Because these communities had difficulties in raising funds to build new treatment plants, the low-cost alternative of constructing wetlands seemed an attractive proposition, but mosquitoes were a problem. From 1988 to 1989, mosquito populations were monitored bi-weekly, and eight of the fourteen species of mosquitoes found breeding in the constructed wetlands had been previously implicated as competent vectors of one or more strains of encephalitis viruses, most notably St Louis encephalitis, which had previously been reported locally. In terms of larval numbers, *Culex pipiens* was often predominant, particularly in the more heavily polluted influent end of the wetlands; this species has been associated with the transmission of St Louis, eastern and western encephalitis viruses, dog heartworm, fowl pox and bird malaria. In adult collections, attack rates of *Culex salinarius* ranged up to 370 per person per 15 minutes, a high rate and intolerable by any standards.

A number of constructed systems has been investigated for biogeochemical efficiency in Australia (e.g. Mitchell *et al.* 1994), but there has been no investigation of production of mosquitoes from such systems. From what is known about mosquito breeding in natural wetlands, we can expect surface flow wetlands and ponded systems with emergent and floating vegetation to present occasional to regular, and sometimes major, mosquito problems.

10.3 Mosquitoes as pests and vectors in Australia

Australia has a history of mosquito problems, pest- and disease-related, and there is evidence that this is increasing in many regions. Humans in Australia have approximately 350 species of mosquitoes to contend with, coupled with about eighty different arboviruses (Russell 1995) and an increasing risk of introduction of exotic arboviruses, regular introduction of malaria, and the unknown threat of exotic emerging diseases. Additionally, we should not forget that mosquitoes and biting midges are associated with arboviruses of veterinary importance, causing congenital defects and ephemeral fever in cattle and blue tongue disease in sheep (Kay and Standfast 1987); mosquitoes are also the vectors of heartworm of dogs, which has high prevalence in many parts of Australia, and of fowl pox.

Much of this increase in mosquito-borne disease has come about because of developments (agricultural, mining, residential and industrial) that have resulted in increased mosquito populations and increased contact between humans and mosquitoes. Neither can do without water. The provision of irrigation to otherwise semi-arid inland areas has brought more than an economic bonus; it has typically resulted in a modified environment, and major increases in the abundance of mosquitoes of nuisance and/or health concern. The construction of wetlands within or in close proximity to residential areas is a concern for individual quality of life and public health; the wetlands may provide habitat not only for mosquitoes, but also for vertebrate reservoirs of mosquito-borne pathogens (e.g. arboviruses) that cause human disease (Russell 1994; 1995).

A number of important mosquitoes colonise wetlands in Australia (Table 10.1). In established sites, species such as *Culex annulirostris*, *Culex quinquefasciatus*, *Coquillettidia linealis*, *Coquillettidia xanthogaster*, *Mansonia uniformis*, *Anopheles bancroftii* and *Anopheles annulipes* might be found, and these are not only important nuisance biters in many areas, but many are potential vectors of disease; others, e.g. *Culex australicus*, *Culex cylindricus*, *Culex globocoxitus*, *Culex hilli*, *Culiseta atra*, *Aedeomyia catasticta*, *Anopheles atratipes* and *Uranotaenia lateralis*, may be abundant but are of little or no significance as pests or vectors of human disease. Periodically flooded wetlands, or those with fluctuations in water level, can be exploited by species such as *Aedes alboannulatus*, *Aedes bancroftianus*, *Aedes clelandi*, *Aedes normanensis*, *Aedes theobaldi*, *Aedes sagax* and *Aedes vittiger*; these mosquitoes are otherwise associated with floodwater pondings and transient wetlands, and can be very troublesome because they bite during the day as well as the evening, and may carry arboviruses (Russell 1994; 1995).

Not all wetlands are fresh water; brackish wetlands of mangrove or paperbark (*Melaleuca*) swamp, and saltmarsh or mudflat depressions, may provide major problems with species such as *Aedes alternans*, *Aedes camptorhynchus*, *Aedes funereus*, *Aedes vigilax*, *Anopheles hilli* and *Culex sitiens*; brackish to saline ponds created for industrial effluent, such as alumina plant effluent in Queensland (Kay *et al.* 1981) and the Northern Territory (Russell *et al.* unpublished data; Whelan *et al.* unpublished data), have produced major mosquito problems associated with the pest species *Aedes vigilax* and *Culex sitiens* in coastal locations.

Mosquitoes are dependent on seasonal conditions and availability of habitat, and there have been various relevant studies in inland and coastal regions of southern (e.g. Russell 1986a; 1986b; Russell *et al.* 1992) and northern Australia (e.g. Barker-Hudson *et al.* 1993; Kay *et al.* 1992; Russell and Whelan 1986; Whelan 1982; Wright 1981). Communities with natural or artificial wetlands have greater populations of mosquitoes than do communities without such habitats, and this is particularly the case in arid and semi-arid regions of Australia (Kay *et al.*, unpublished observations).

Although there is little experience of control of mosquitoes in wetlands in Australia, the biology of the most important species has been investigated in a number of localities. The seasonal activity and relative abundance of mosquitoes in various areas of Australia, particularly for the major pest species on the coast (*Aedes camptorhynchus*, *Aedes vigilax*, *Culex sitiens*) and inland (*Aedes theobaldi*, *Aedes vittiger*, *Anopheles annulipes*, *Anopheles bancroftii*, *Culex annulirostris*), and other important species (e.g. *Aedes normanensis*, *Aedes notoscriptus*, *Coquillettidia linealis*, *Coquillettidia xanthogaster*, *Mansonia uniformis*), is now known about in some detail. The monograph series *The Culicidae of the Australasian Region* produced in 12 volumes between 1980 and 1989 by Lee *et al.* (1980) provides a wealth of information on the distribution, biology and relation to disease of Australasian mosquitoes.

Other aspects of mosquito biology important to pest and vector activity, including feeding behaviour, longevity and flight range, have also been studied. These data allow informed decisions on control of mosquito populations and protection of human communities. For instance, information on effective flight range of important species can be used to determine the preferred distance for siting constructed wetlands away from residential areas; we know, for instance, that *Culex annulirostris* is able to disperse 5–10 km in south-eastern Australia (Bryan *et al.* 1992; Russell 1986c).

Culex annulirostris in Australian wetlands is the ecological counterpart of *Culex tarsalis* in North America and *Culex tritaeniorhynchus* in east Asia with respect to biology and status as major pests and vectors of disease. In northern Australia, *Culex annulirostris* may be active throughout the year, but is often most abundant towards the end of, or shortly after, the wet season. In southern

Table 10.1 Mosquitoes found in wetlands in Australia, their pest/vector status and associated pathogens

Mosquito species[a]	*Distribution*[b]	*Wetland type*	*Pest/vector status*[c]
Anopheles annulipes	All states/territories	Fresh water	Pest; virus (BF, RR, MVE); malaria; heartworm
Anopheles atratipes	NSW,QLD,SA,TAS,VIC,WA	Fresh water	Non-pest
Anopheles bancroftii	NT,QLD,WA	Fresh water	Pest; virus (MVE, BEF); malaria
Anopheles farauti	QLD, NT, WA	Brackish/fresh	Minor pest; malaria
Anopheles hilli	NT,QLD,WA	Brackish	Pest; malaria
Anopheles meraukensis	NT,QLD,WA	Fresh water	Minor pest
Aedeomyia catasticta	NT,QLD,WA	Fresh water (filamentous algae)	Non-pest
Aedeomyia venustipes	NSW,QLD,VIC,TAS	Fresh water (filamentous algae)	Non-pest
Aedes alboannulatus	NSW,QLD,SA,TAS,VIC,WA	Fresh water	Pest
Aedes alternans	NSW,NT,QLD,SA,VIC,WA	Saline/brackish/fresh	Pest; virus (RR)
Aedes bancroftianus	NSW,NT,QLD,SA,VIC,WA	Fresh water, transient, flood	Pest; virus (BF, RR)
Aedes camptorhynchus	NSW,SA,TAS,VIC,WA	Saline/brackish/fresh	Pest; virus (BF, RR); heartworm
Aedes clelandi	SA,TAS,VIC,WA	Fresh water	Pest; virus (RR)
Aedes flavifrons	NSW,SA,TAS,VIC	Brackish/fresh	Pest; virus (RR)
Aedes funereus	NSW,NT,QLD,	Brackish/fresh	Pest; virus (BF, RR)
Aedes procax	NSW,QLD,VIC	Fresh water, transient, flood	Pest; virus (BF, RR)
Aedes normanensis	NSW,NT,QLD,WA	Fresh water, transient, flood	Pest; virus (BF, MVE, RR)
Aedes sagax	NSW,QLD,SA,VIC,WA	Fresh water, transient, flood	Pest; virus (RR)
Aedes reesi	NT,QLD,WA	Fresh water	Pest
Aedes theobaldi	NSW,QLD,SA,VIC	Fresh water, transient, flood	Pest; virus (RR)
Aedes vigilax	NSW,NT,QLD,SA,WA	Saline/brackish, transient	Pest; virus (BF, RR); heartworm
Aedes vittiger	NSW,NT,QLD,SA,VIC	Fresh water, transient, flood	Pest

Table 10.1 continued

Mosquito species[a]	*Distribution*[b]	*Wetland type*	*Pest/vector status*[c]
Coquillettidia linealis	NSW,QLD,SA,TAS,VIC,WA	Fresh water (emergent vegetation)	Pest; virus (BF, RR)
Coq. variegata	NSW,QLD,VIC	Fresh water (emergent vegetation)	Pest
Coq. xanthogaster	NSW,NT,QLD,WA	Fresh water (emergent vegetation)	Pest
Culex annulirostris	All states/territories (? TAS)	Fresh water	Pest; virus (BF, JE, KUN, MVE, RR, BEF); heartworm
Culex australicus	All states/territories	Fresh water	Non-pest; virus (KUN, MVE, RR)
Culex bitaeniorhynchus	NSW,NT,QLD,WA	Fresh water	Pest; virus
Culex cylindricus	NSW,NT,QLD,VIC	Fresh water	Non-pest
Culex globocoxitus	All states/territories	Brackish/ fresh	Non-pest
Culex hilli	NT,QLD,WA	Fresh water	Non-pest
Culex palpalis	NSW,NT,QLD,WA	Fresh water	Minor pest
Culex pullus	NT,QLD,WA	Fresh water	Non-pest
Culex quinquefasciatus	All states/territories	Fresh water (polluted)	Pest; virus (BF, KUN, MVE, RR); heartworm; fowl pox
Culex sitiens	NSW,NT,QLD,WA	Saline/brackish	Pest; virus (BF, RR)
Culiseta atra	WA	Fresh water	Non-pest
Mansonia uniformis	NSW,NT,QLD,VIC,WA	Fresh water (floating vegetation)	Pest; virus (MVE, RR)
Uranotaenia lateralis	NSW,NT,QLD	Fresh water	Non-pest

[a] For details see: Dobrotworsky (1965); Kay *et al.* (1981); Lee *et al.* (1980); Liehne (1991); Marks (1982); Russell (1993; 1995); Whelan (1982; 1984).
[b] NSW, New South Wales; NT, Northern Territory; QLD, Queensland; SA, South Australia; TAS, Tasmania; VIC, Victoria; WA, Western Australia.
[c] Arboviruses: BEF, bovine ephemeral fever; BF, Barmah Forest; JE, Japanese encephalitis; KUN, Kunjin; MVE, Murray Valley encephalitis; RR, Ross River.

Australia, *Culex annulirostris* is seasonal; it disappears during winter but is dominant during mid- to late summer. The species can colonize shallow vegetated fresh water pools within one day of formation, and most larvae are associated with the vegetation; maximum densities may be attained in about a week, but densities may then gradually decline as the habitat becomes stabilized with a greater range of fauna, including other insects that prey on mosquitoes (McDonald and Buchanan 1981).

The generally abundant summer pest mosquito in inland area, *Culex annulirostris*, is the major vector of the life-threatening Murray Valley encephalitis (MVE) and Kunjin flaviviruses, the debilitating arthritis-causing Ross River (RR) and Barmah Forest (BF) alphaviruses, and other arboviruses in Australia (Russell 1993; 1994; 1995). This mosquito was also associated with Japanese encephalitis (JE) virus in the recent outbreak in the Torres Strait islands, and if JE virus was to invade and become established on the mainland, *Culex annulirostris* could become involved in the transmission of this life-threatening virus over a very large area of Australia, north and south.

MVE, Kunjin and JE viruses are transmitted from waterbird hosts, and activity of these viruses is associated with wetlands where the birds congregate in close conjunction with mosquitoes, especially *Culex annulirostris*. MVE virus has not been active in southern Australia for some years but infections occur more often in northern regions (see Chapter 8); Kunjin virus has been active in the north and south-east of Australia in recent decades, and JE appears to be a new and distinct threat to the north. Recently, West Nile virus was reported as another immigrant into north-west Western Australia.

RR virus is seasonally active each year throughout Australia and is a major public health concern; in recent years it has caused an average of several thousand confirmed cases annually (Russell 1994; 1995). Many mosquitoes have been incriminated as vectors: *Culex annulirostris* and *Coquillettidia linealis* are likely to be two of the most important summer vectors for inland fresh water wetlands, although various *Aedes* species are involved opportunistically; and in coastal regions *Aedes vigilax* and *Aedes camptorhynchus* from saline wetlands appear the major vectors. BF virus is also widely distributed in Australia and its activity appears to be increasing; its ecology seems to be similar to that of RR virus although little is known about it and its vectors.

Studies on human residents of all regions of NSW during the early 1980s showed the western areas of the state to have the highest levels (30–40+ per cent prevalence) of infection with arboviruses. Ten years later, a follow-up investigation demonstrated increasing arboviral infections, and evidence that a number of as yet unrecognized arboviruses are infecting humans. This situation may be paralleled in other states.

And what of global warming and the 'greenhouse effect'? If predictions (and there are many) are true, increased winter temperatures and summer rainfall will extend the season and habitat of the above-mentioned species, and increase associated pest problems and disease risks. Warmer temperatures will hasten development of aquatic mosquito stages; thus there will be a smaller 'window' for control. Warmer temperatures usually decrease the incubation periods of pathogens in mosquitoes; thus mosquitoes will not have to live as long before they become dangerous. The provision of constructed wetlands as additional habitats within flight range of residential areas will increase mosquito populations and the opportunities for virus transmission. This can be guaranteed to provide a concern for public health unless the mosquitoes are controlled with appropriate and effective management programmes.

10.4 Mosquito monitoring in wetlands

It is not always easy to determine the level at which mosquitoes become a problem. A 'threshold level' will usually be site-specific, and a 'hazard' will be determined by the species, its disease potential and pest significance. A risk assessment will include consideration of relative abundance, flight range and distance from human habitation, and the exposure of the human population that will occur through occupational and recreational activities. A programme for surveillance of the mosquito populations, monitoring the various species and their relative abundance, will provide the necessary data.

Mosquito populations can be monitored by regular (at least weekly) inspections for larvae, and with collections using standard dippers or nets. This will give an indication of the species present and their relative abundance between sampling periods. However, an assessment of the full extent of mosquito production might only be possible with regular thorough larval surveys using techniques involving replicated dip-sampling, at pre-selected points around the wetland and on transects within it; this repetitive sampling should be undertaken with concurrent adult monitoring to compare samples from the site and from other parts of the surrounding and nearby areas. Adult monitoring requires periodic sampling with standard traps (preferably a dry ice-baited encephalitis virus surveillance type (Figure 10.1), such as that sold by Australian Entomological Supplies), at and near the site, in concert with simultaneous collections elsewhere in the area to provide reference data for comparison. In general:

- larval collections indicate the species in the wetlands, and their relative abundance
- adult collections confirm the species in the area, and the pests or vectors present
- larval sampling indicates the suitability of the habitat
- adult sampling indicates the productivity of the habitat
- larval sampling may indicate an abundance of immatures that may not translate into an abundance of adults because of various factors, e.g. predation
- adult sampling may collect mosquitoes that are not being produced by the habitat in question, i.e. they are flying in from elsewhere
- therefore, both larval and adult monitoring are essential.

Figure 10.1 Mosquitoes and other blood sucking flies are attracted to encephalitis virus surveillance traps via a small light and carbon dioxide emanated by dry ice. A battery-operated fan provides the downdraft to trap them in the bag.

10.5 Managing mosquito populations

Although water treatment and wildlife habitat may be the priorities for constructed wetlands, the objective of mosquito management should require that mosquito prevention design, water and vegetation management, and the use of mosquito control agents (chemical and biological) be considered in the planning, operation and maintenance of wetlands. This concept of 'integrated control' against mosquitoes has been well illustrated in Australia in the city and suburbs of Darwin (Whelan 1989).

10.5.1 Design and construction influences

Constructed wetlands may contain various areas designed to cleanse, treat and polish the inflowing water in turn. Sedimentation ponds intended to be 'deep', and vegetated wetlands intended to be 'shallow', provide different opportunities for mosquitoes. Inflow water that is primarily storm water or agricultural run-off, not heavily polluted with organic material, will usually create less of a problem than sewage or similar wastewater. In many wetlands, high nutrient content of the inflow will enhance vegetation and mosquito production, although other pollutants (e.g. chemical) can be detrimental.

In large ponds, it is advisable to profile the bottom so that the depth is greatest at the inflow end, particularly if there is seasonal variation in input. However, if the site is too large relative to the expected inflow and required deposition of solids, the pond may quickly become a shallow, thickly vegetated marsh capable of supporting very large numbers of mosquitoes.

Retention ponds are unlikely to provide much opportunity for mosquito breeding in deeper (approximately 2 m) areas of open water, so the concern is primarily for the margins. For minimum mosquito potential, the margins should be a vertical concrete design; however, if this is inappropriate, the margins should be kept as steep, deep and clean as possible.

Ideally, to control mosquitoes, wetlands should be located in open areas where wind action is detrimental to mosquito survival and may also prevent the growth of floating algae, aquatic ferns (e.g. *Azolla* sp.), and plants such as duckweed (*Lemna* sp.), which may provide protection for mosquito larvae from wave action and predators.

In general, the construction options for wetland impoundments to prevent emergent vegetation contributing to mosquito problems are a steep (>30°) edge and a deep (2–2.5 m) bottom. However, this may not be an acceptable condition where there is concern for human safety or requirements for a shallow habitat for wading birds.

10.5.2 Vegetation influences

Most wetlands that do not support vegetation also do not support large populations of mosquitoes, at least in the long term. Mosquito populations that colonize such wetlands are suppressed (but not necessarily eliminated) by predation, physical disturbance, or depletion of food resources. Vegetation provides protection for mosquitoes from predation and physical disturbance, and oviposition substrates and favourable thermal conditions for rapid development. Vegetation also provides enhanced food resources, and although this varies with different mosquitoes and plants, more complex systems can provide more nutrients. The factors determining mosquito populations, even in monoculture wetlands, are complex and vary from site to site and throughout a season.

Plants that might colonize wetlands in Australia include *Myriophyllum*, *Potamogeton*, *Eleocharis*, *Juncus*, *Bolboschoenus*, *Schoenoplectus*, *Typha* and

Phragmites, and various species of algae (Sainty and Jacobs 1981); all can promote mosquito colonization. *Culex* mosquito production is primarily associated with edge vegetation; floating vegetation is generally more important for *Anopheles* mosquitoes but is rarely important for *Aedes* mosquitoes. Vegetation that floats on the water, but does not completely occlude it, may support a diverse mosquito fauna of *Anopheles* and *Culex* species. Dense patches of free-floating vascular plants (e.g. the waterfern *Azolla*, the duckweeds *Lemna* and *Spirodela*) that completely cover the surface may inhibit egg laying and pupal emergence for some mosquitoes (Rajendran and Reuben 1991), but there are also reports of larval density being highest in ponds with the heaviest duckweed cover (Tennessen and Painter 1990). Reeds, sedges and rushes (e.g. *Typha*, *Phragmites*, *Schoenoplectus*, *Bolboschoenus*) often produce major mosquito problems when the plants becomes extremely dense, and when they die and/or become lodged so that predators cannot gain access to mosquito larvae.

The options for the management of plants include maintenance, reduction, or elimination by harvesting, water regulation, herbicide application, and herbivore introduction (herbivorous waterfowl can indirectly reduce mosquito populations by reducing refuge and increasing open areas for physical disturbance and access by fish and other predators). Although colonization of wetlands by invertebrate predators is usually governed by natural processes that cannot be regulated, predation by vertebrate and invertebrate predators can be aided by vegetation management. In the absence of fish, a relative abundance of some predatory insects, e.g. backswimmers (Notonectidae), has increased with the removal of emergent and submergent vegetation; in contrast, the addition of short emergent or submergent vegetation has increased the abundance of dragonflies and damselflies (Collins and Resh 1989).

10.5.3 Water management

The control of mosquitoes (and vegetation) by water level regulation, particularly through periodic drainage, is an option for many wetlands. *Culex* and *Anopheles* populations can be controlled by periodic drainage, carefully arranged so as not to adversely affect marginal and emergent vegetation but timed so that regimes are shorter than the developmental period of the larval stages of the particular target mosquito species; however, such measures may increase populations of certain *Aedes* species and the local situation must be considered.

Marginal and floating vegetation can sometimes be controlled by water level management, alternately flooding and stranding the plants, but this does little for non-marginal vegetation unless the site is completely dried; even then it may not be successful.

Increasing depth in shallow habitats has reduced emergent vegetation and mosquito production in some situations; in wetlands with thick vegetation in California, deeper (60 cm) waters had lower densities of mosquitoes and higher densities of other invertebrates than shallower (20 cm and 40 cm) waters (Batzer and Resh 1992). However, while *Typha* is often thought to be unable to survive in water that is more than 1 m in depth, *Typha* and *Eleocharis* spp. can grow in habitats up to 2 m deep in Australia (Sainty and Jacobs 1981) and, under such circumstances, mosquito populations may persist.

In California, a novel approach, using a sprinkler system to inhibit mosquito oviposition in constructed wetlands, has proven effective in reducing *Culex* populations in a water hyacinth (*Eichhornia crassipes*) system in San Diego (Ebipane *et al.* 1993), and in a *Scirpus acutus/Typha latifolia* system in Sacramento (Wright, personal communication).

10.5.4 Mosquito control agents

If surveillance indicates a problem (or a potential one), what do you do in order to reduce a pest nuisance or remove a threat of disease? There are a few options:

Chemical methods

These have the principal advantage that they can be organized and employed within a short period of time and produce quick results at relatively low cost. However, there is a relatively small number of chemical agents available for mosquito control, and their effectiveness depends on the nature of the habitat, the environmental conditions, and the availability of an appropriate formulation. Agents that are often recommended are listed below.

- One of the commercial preparations of the toxic products of the bacterium *Bacillus thuringiensis israelensis*, commonly known as *Bti* (e.g. Skeetal®, Vectobac®, Cybate®). These are particularly environmentally acceptable because of an almost total specificity for mosquitoes among invertebrates, and a negligible toxicity for vertebrates. The agent has had widespread use in Australia. Weekly applications may be required because there is little persistence and no recycling. A similar agent made from *Bacillus sphaericus* has recently become commercially available in the USA; the product, Vectolex®, is designed for wetlands with polluted water and species such as *Culex quinquefasciatus*.
- The insect growth regulator methoprene (Altosid®) is a juvenile hormone mimic chemical that is environmentally efficacious because of its relative specificity for mosquitoes among aquatic animals and its negligible toxicity for vertebrates. Various formulations are available, and some slow-release products may provide control for three to six months. It can be difficult to assess the effectiveness of this chemical by larval surveys only, because it does not kill the larvae; it simply stops development to the adult mosquito. Thus the monitoring of efficacy is done by inspecting caged immature stages in the field.
- The organophosphate temephos (Abate®), a pesticide that is toxic to

invertebrates, has been the mainstay of Australian mosquito control for over twenty years. It can be toxic to fish and birds, but is safe for human use; it has a particular disadvantage in its deleterious effects on some crustaceans, and thus newer products are being preferred. It is relatively target-specific for mosquitoes at recommended dosages, and has been commonly used in environmentally sensitive areas. Its persistence is limited and weekly applications might be required to control breeding in fresh water wetlands during summer.

Chemical control is only a stop-gap strategy but should be used to manage unforeseen episodes of heavy breeding. Neglect of design aspects at the outset could mean indeterminate recurrent expenditure on maintenance. Additionally, the potential impact of biological factors, e.g. natural predation through fish, some of the aquatic beetles or bugs, or other predators or pathogens could influence the frequency of chemical interventions.

Biological methods

These appeal as an attractive alternative but, with the exception of fish, no biological agents are available for practical use in wetlands. Larvivorous fish can be an extremely valuable component of an integrated control programme, but in shallow wetlands the type and density of vegetation are critical in determining their effectiveness; pollution can also be a limiting factor that must be considered (Mian *et al.* 1986). Sink basins can be useful as reservoirs to preserve predator fish populations through dry times.

Indigenous species should be given priority, although the exotic mosquito fish *Gambusia holbrooki* (previously known as *G. affinis*) may be difficult to exclude if it is present (as it is widely) in regional drainage systems and local watercourses. *Gambusia holbrooki* can be an effective mosquito control agent but its role in Australia is controversial; it has adverse effects on local fauna, and there are native fish that might be preferred (Lloyd 1986).

Discussion with experts familiar with local species would be appropriate but, whichever fish is used, experience indicates that it should be introduced in substantial numbers. To maintain fish predation, vegetation may have to be periodically removed or harvested; water level variation can be used to discourage certain plants, as mentioned above, and spatial arrangement of vegetation types can be used to maximize fish foraging patterns.

10.6 Conclusions and general recommendations

Constructed wetlands provide for water conservation, protection of public health and environmental resources. As the extent and quality of natural wetlands decreases, the value of remaining wildlife habitats and that which can be provided by artificial wetlands increases. Improved understanding of the complex interrelated process mechanisms associated with different types of wetland habitats will contribute to wetland design that optimizes the efficiency of objectives related to both water purification and maintenance of natural ecosystems (Stiles 1994). Moreover, there is a need for research into mosquito production from, and mosquito control in, constructed wetlands in Australia. Mosquito problems vary with region and type of wetland, and investigations into mosquito production should be included in the monitoring of a range of wetland developments.

Any wetland area constructed as a shallow pond that is extensively to completely vegetated has the potential for mosquito breeding. Deep ponds with steep edges and no emergent or surface vegetation provide significantly less habitat for mosquitoes. Design of artificial wetlands to militate against mosquitoes is important because 'prevention is better than cure'. It costs less in the end, not just in fiscal terms but also in terms of human suffering. Emergency mosquito control with chemical or biological control agents can be very difficult in wastewater treatment facilities. Insecticides that work well in unpolluted habitats can be considerably less effective in wastewaters because of binding to organic matter. Also, predatory fish and other insects will not tolerate the low oxygen conditions that prevail in some treatment facilities and will exert less than expected control.

Most constructed wetlands will occasionally require a programme of mosquito prevention and control designed for the particular local circumstances, and the programme should consist of an integrated variety of control approaches. When chemical or biological agents do not adversely affect mosquito predators and pathogens, season-long mosquito control may be possible after only one or two treatments. Highly selective mosquito larvicides should be used strategically at optimal dosages when mosquito populations require intervention; this requires frequent surveillance of the wetland and careful monitoring of the mosquito breeding activity. Mosquito control should be a consideration in the planning, operation and management phases for all constructed wetland systems treating effluent waters.

The design of constructed wetlands in Australia has been receiving attention in recent years and, within concept designs, the various constraints associated with wetlands are being considered in greater detail; undesirable functions such as the ability of wetland systems to produce mosquitoes are no longer being ignored and are being included in process models such as that developed by the NSW Department of Land and Water Conservation (Russell and Kuginis 1998; White 1995; DLWC 1998), and by the Queensland Department of Primary Industries (Simpson 1995). One wetland has been constructed in NSW with a design to create, as far as possible, conditions that are unattractive to mosquito development (Russell and Kuginis 1995). The effectiveness of the design criteria and operational considerations will become apparent only after some years of monitoring, but they follow those discussed in detail in section 10.5 above, and are summarized below.

- Ponds should be deep, preferably 2.5 m in the centre, with edges sloping at not less than 3:1, and the margin maintained free of vegetation. These ponds should have a regular perimeter; accumulations of debris, vegetation and algal mats should be routinely removed from the water surface.

- Ponds that are necessarily shallow should be graded for rapid dewatering; weirs, siphons and pumps can be provided for water level management. These ponds should be maintained with steep edges, free of emergent vegetation, and the interior of the pond should be maintained free of emergent and floating vegetation.
- Ponds that are necessarily vegetated and shallow should be maintained to reduce mosquito populations with water management practices such as alternate flooding and draining. Areas of deep water should be provided to enable fish to survive dewatering periods. The duration of 'dry' time should be discussed with plant and mosquito experts.
- Ponds that are necessarily shallow and vegetated which cannot be periodically drained to eliminate mosquito populations should be maintained with populations of efficient indigenous vertebrate (fish) and invertebrate (e.g. water beetles, water bugs, dragonflies) predators; sprinkler systems may be useful to hinder mosquito colonization.
- Ponds in any of the above categories should be regularly checked with standard mosquito sampling techniques to detect mosquito breeding and monitor the relative abundance of both larval and adult populations associated with the habitat.
- Ponds that are found to be producing unacceptably high levels of mosquitoes and that cannot be 'treated' with any structural, environmental or biological methods should be treated by judicious application of a recommended chemical control agent.

References

Barker-Hudson, P., Kay, B.H., Jones, R.E. *et al.* (1993) Surveillance of mosquitoes and arbovirus infection at the Ross River dam (Stage 1), Australia. *Journal of the American Mosquito Control Association*, **9**, 389–394.

Batzer, D.P. and Resh, V.H. (1992) Wetland management strategies that enhance waterfowl habitats can also control mosquitoes. *Journal of the American Mosquito Control Association*, **8**, 117–125.

Breen, P.F. (1989) Hydrology, structure and function of natural wetlands. In: *Proceedings Applied Ecology and Conservation Seminar Series*. La Trobe University, Melbourne, pp. 31–38.

Bryan, J.H., O'Donnell, M.S., Berry, G. *et al.* (1992) Dispersal of adult female *Culex annulirostris* in Griffith, New South Wales, Australia: a further study. *Journal of the American Mosquito Control Association*, **8**, 398–403.

Collins, J.N. and Resh, V.H. (1989) *Guidelines for the ecological control of mosquitoes in non-tidal wetlands of the San Francisco Bay Area*. California Mosquito and Vector Control Association and The University of California Mosquito Research Program, Berkeley, California.

Dinges, R. (1978) Aquatic vegetation and water pollution control public health implications. *American Journal of Public Health*, **68**, 1202–1205.

DLWC (1998) *The Constructed Wetlands Manual.* Dept Land and Water Conservation, Sydney, Orbital Offset Printers, 424 pp.

Dobrotworsky, N.V. (1965) *The Mosquitoes of Victoria.* Melbourne University Press, Melbourne.

Dunkerley, G.M. and Carter, R.M. (1992) Wetland management – the NSW wetlands policy. In: *Wetlands down under – wetland systems in water pollution control, Proceedings of a conference.* University of New South Wales, Sydney, Australia, pp. 58.1–58.7.

Ebipane, R., Heidig, E. and Gibson, D.W. (1993) Prevention of mosquito production at an aquaculture wastewater reclamation plant in San Diego using an innovative sprinkler system. *Bulletin of the Society for Vector Ecology*, **18**, 40–44.

Eldridge, B.F. and Martin, C. (1987) Mosquito problems in sewage treatment plants using aquatic macrophages in California. *Proceedings and Papers of the Annual Conference of the California Mosquito and Vector Control Association*, **55**, 87–91.

Finlayson, C.M. and Chick, A.J. (1983) Testing the potential of aquatic plants to treat abattoir effluent. *Water Research*, **17**, 415–422.

Hammer, D.A. (1992) *Creating freshwater wetlands.* Lewis Publishers, Boca Raton, USA.

Kay, B.H. and Standfast, H.A. (1987) Ecology of arboviruses and their vectors in Australia, *Current Topics in Vector Research*, **3**, 1–36.

Kay, B.H., Piper, R.G., Falk, P.E. *et al.* (1992) Mosquitoes from ricefields at Mareeba, North Queensland, Australia. *General and Applied Entomology*, **24**, 19–32.

Kay, B.H., Sinclair, P. and Marks, E.N. (1981) Mosquitoes: their interrelationships with man. In: *The Ecology of Pests.* (eds R.L. Kitching and R.E. Jones), CSIRO, Melbourne, pp. 157–174.

Kramer, V.L. and Garcia, R. (1989) An analysis of factors affecting mosquito abundance in California wild rice fields, *Bulletin of the Society for Vector Ecology*, **14**, 87–92.

Lacey, L.A. and Lacey, C.M. (1990) The medical importance of riceland mosquitoes and their control using alternatives to chemical insecticides. *Journal of the American Mosquito Control Association*, **6** (Supplement #2), 1–93.

Lee, D.J., Hicks, M.M., Griffiths, M., Russell, R.C. and Marks, E.N. (1980) *The Culicidae of the Australasian Region*, Volume 1. Australian Government Publishing Service, Canberra. (Volumes 2–12 published 1982 through 1989.)

Liehne, P.F.S. (1991) *An Atlas of the Mosquitoes of Western Australia.* Health Department of Western Australia, Perth.

Lloyd, L. (1986) An alternative to insect control by 'mosquitofish', *Gambusia affinis. Arbovirus Research in Australia*, **4**, 156–163.

McDonald, G. and Buchanan G.A. (1981) The mosquito and predatory insect fauna inhabiting fresh-water ponds, with particular reference to *Culex annulirostris* Skuse (Diptera: Culicidae). *Australian Journal of Ecology*, **6**, 21–27.

Marks, E.N. (1982) *An Atlas of Common Queensland Mosquitoes.* Queensland Institute of Medical Research, Brisbane.

Martin, C.V. and Eldridge, B.F. (1989) California's experience with mosquitoes in aquatic wastewater treatment systems. In: *Constructed wetlands for wastewater treatment: municipal, industrial and agricultural* (ed. D.A. Hammer). Lewis Publishers, Michigan, USA, pp. 393–398.

Mian, L.S., Mulla, M.S. and Wilson, B.A. (1986) Studies on potential biological control agents of immature mosquitoes in sewage wastewater in southern California. *Journal of the American Mosquito Control Association*, **2**, 329–325.

Mitchell, D.S., Chick, A.J. and Raisin, G.W. (1994) The use of wetlands for water pollution control in Australia: an ecological perspective. In: *Proceedings 4th International Conference on Wetland Systems for Water Pollution Control*. Guangzhou, China, pp. 709–716.

Mortenson, E.W. (1983) Mosquito occurrence in wastewater marshes: a potential new community problem. *Proceedings of the California Mosquito and Vector Control Association*, **50**, 65–67.

Rajendran, R. and Reuben, R. (1991) Evaluation of the water fern *Azolla microphylla* for mosquito population management in the rice-land agro-ecosystem of south India. *Medical and Veterinary Entomology*, **5**, 299–310.

Roser, D.J. and Bavor, H.J. (1994) Current directions in constructed wetlands research in New South Wales. In: *Practical Aspects of Constructed Wetlands, Proceedings of a workshop at Shortland Wetlands Centre*. NSW Department of Water Resources, Sydney, pp. 53–57.

Russell, R.C. (1986a) Seasonal abundance of mosquitoes in a native forest of the Murray Valley of Victoria, 1979–1985. *Journal of the Australian Entomological Society*, **25**, 235–240.

Russell, R.C. (1986b) The mosquito fauna of Conjola State Forest on the south coast of New South Wales. Part 1. Species composition and monthly prevalence. *General and Applied Entomology*, **18**, 53–64.

Russell, R.C. (1986c) Dispersal of the arbovirus vector *Culex annulirostris* Skuse (Diptera: Culicidae) in the Murray Valley of Victoria, Australia. *General and Applied Entomology*, **18**, 5–9.

Russell, R.C. (1993) *Mosquitoes and mosquito-borne disease in southeastern Australia*, University of Sydney, Department of Medical Entomology, Westmead Hospital, Westmead.

Russell, R.C. (1994) Wetlands and mosquitoes – the pest and disease threats. In: *Practical Aspects of Constructed Wetlands, Proceedings of a workshop at Shortland Wetlands Centre*. NSW Department of Water Resources, Sydney, pp. 45–51.

Russell, R.C. (1995) Arboviruses and their vectors in Australia: an update on the ecology and epidemiology of some mosquito-borne arboviruses. *Review of Medical and Veterinary Entomology*, **83**, 141–158.

Russell, R.C. and Kuginis, L. (1996) Mosquito Risk Assessment and Management. In: *The Constructed Wetlands Manual* (ed. DLWC). NSW Department of Land and Water Conservation, Sydney, pp. 181–191.

Russell, R.C. and Kuginis, L. (1995) Constructed wetlands and mosquitoes – some problems and some solutions. In: *Wetlands for Water Quality Control. Proceedings of a National Conference*. Queensland Department of Primary Industries, Brisbane, pp. 213–224.

Russell, R.C. and Whelan, P.I. (1986) Seasonal prevalence of adult mosquitoes at Casuarina and Leanyer, Darwin. *Australian Journal of Ecology*, **11**, 99–105.

Russell, R.C., Vale, T.G., Wells, P.J. *et al.* (1992) Monthly periodicity and abundance of mosquitoes (Diptera: Culicidae) near Batemans Bay on the south coast of New South Wales, 1985–1988. *Journal of the Australian Entomological Society*, **31**, 281–288.

Sainty, G.R. and Jacobs, S.W.L. (1981) *Waterplants of New South Wales*. Water Resources Commission NSW, Sydney.

Schaefer, C.H., Miura, T., Stewart, R.J. *et al.* (1983) Studies on the relationship of mosquito breeding on rice fields and the use of sewage effluent for irrigation. *Proceedings of the California Mosquito and Vector Control Association*, **50**, 59–65.

Simpson, J.S. (1995) *Interim Guidelines on the Planning, Design and Management of artificial wetlands in Queensland*, Queensland Department of Primary Industries, Brisbane.

Stiles, E.A. (1994) An integrated design approach for multipurpose wetlands. In: *Proceedings 4th International Conference on Wetland Systems for Water Pollution Control.* Guangzhou, China, pp. 723–731.

Tennessen, K.J. and Painter, M.K. (1990) *Mosquito Production in Constructed Wetlands for Treatment of Municipal Wastewater.* Tennessee Valley Authority, Technical report TVA/WR/AB–90/4.

Thomas, P.R., Glover, P. and Kalaroopan, T. (1994) An evaluation of pollutant removal from secondary treated sewage effluent using a constructed wetland system. In: *Proceedings 4th International Conference on Wetland Systems for Water Pollution Control.* Guangzhou, China, pp. 421–427.

Verhoeven, T.J. and White, G.C. (1994) Belubula catchment constructed ecosystems New South Wales nutrient control. In: *Proceedings 4th International Conference on Wetland Systems for Water Pollution Control.* Guangzhou, China, pp. 654–662.

Whelan, P.I. (1982) *Mosquito Breeding in Darwin.* Northern Territory Department of Health, Darwin.

Whelan, P.I. (1984) *Mosquitoes of Public Health Importance in the Northern Territory and their Control.* Northern Territory Department of Health, Darwin.

Whelan, P.I. (1989) Integrated mosquito control in Darwin. *Arbovirus Research in Australia*, **5**, 178–185.

White, G. (1995) The design of constructed wetlands in New South Wales. In: *Wetlands for Water Quality Control, Proceedings of a National Conference.* Queensland Department of Primary Industries, Brisbane, pp. 425–434.

Wright, A.E. (1981) Ord River arboviruses – mosquito captures during 1976/77. *Journal of the Australian Entomological Society*, **20**, 47–57.

Chapter 11

Environmental concerns and recreational development in relation to public health issues associated with water resource projects of the Tennessee Valley Authority*

Joseph C. Cooney†, Kenneth J. Tennessen, and Edward L. Snoddy (deceased)

11.1 Introduction

Environmental considerations and recreational development are major issues that must be addressed in assessing the need for and value of water resource projects. In considering proposals for these types of projects for approval and potential for funding, attention should be given to the following four major topic areas:

- institutional arrangements
- project purpose and priorities of operation
- public health considerations
- environmental regulations and compliance.

A description of the elements contained in these topic areas is provided below, followed by a detailed discussion of the experience of the Tennessee Valley Authority (TVA) in dealing with these issues.

Institutional arrangements (the organization, its membership, and functioning authority) must allow for a broad range of project considerations in the planning and developmental phases as well as during operation of the project. Project management should include expertise in disciplines that can address diverse user interests, such as environmental engineering, public health and vector ecology, fish biology, water quality, aquatic ecology, etc. Management should have the authority to make decisions and authorize changes in operations to accommodate environmental and political issues as well as customer demands.

Water resource development projects can be proposed for single purposes, such as water supply, wastewater processing, hydropower, flood control, navigation,

* The views expressed in this chapter reflect those of the authors and are not current TVA policy.

† Joseph Cooney is independent of the TVA.

irrigation, or recreation. Any of these purposes can be combined with each other or with other purposes such as shoreline development, fish and wildlife management, public health, or a number of others to create multipurpose projects. Priorities for operation and use must be established and adhered to, otherwise use conflicts will occur and optimum benefits will not accrue. Regardless of the project purpose, other use requests will evolve as other potentials are realized; consequently, it is imperative that primary purposes are carefully documented and maintained. Regardless of the purpose, if static water is allowed to pond and is exposed to the natural environment, public health as it relates to the transmission of mosquito-borne diseases is jeopardized and this issue must be addressed in the planning, developmental, and operational phases of the project.

Public health considerations should be inherent in the planning, construction, and operation of water resource projects, especially if they are multipurpose in nature encouraging human use and water contact, because health problems can be expected when water is impounded. Public health issues range from basic water quality problems to problems caused by water-borne pathogens and vector-borne diseases. Prior to impounding the Tennessee River, malaria was a major concern of the TVA in the valley. This chapter outlines how TVA worked to eliminate this problem and continues to manage the river system so that disease transmission does not recur and the annoyance level from mosquitoes breeding on the reservoirs is minimized. Mosquito control activities, which benefit the multipurpose scheme and allow for a wide array of recreational uses, are discussed in the following pages.

Environmental regulations that impact the development and use of water resource projects continue to be issued, and compliance is a requirement for both establishment and continued operation of a project. In the USA, federal regulations that directly impact these types of projects and govern their continued operation include:

- the Clean Water Act
- the Endangered Species Act
- wetlands preservation initiatives and executive orders such as the 'No Net Loss of Wetlands' policy.

In addition to these primary regulations, others exist that have secondary or tangential impact, as well as those produced by state governments that require compliance. Continuous monitoring of regulatory changes is necessary so that required operational adjustments can be made to ensure compliance.

11.2 Organization and institutional arrangements

The TVA was created as a regional development agency in 1933 by an Act of the US Congress. It was one element in a broad programme designed to bring the nation out of severe economic troubles. Its structure, however, made it unique among federal agencies. The TVA was set up to function as a government-owned corporation with a three-member board of directors. At the same time, it operates with a reasonable degree of autonomy and the flexibility of a private corporation. Each director is appointed by the US president for a term of up to nine years. The directors' terms are fixed and staggered at three year intervals, so that a director might serve all or only part of a term depending on political and other circumstances.

The TVA is an independent agency, and not part of any federal cabinet department. Consequently, interdepartmental conflicts are limited. The TVA Act provided the agency with administrative freedom to meet the special requirements of its programme and to adopt the methods of administration of successful private as well as public enterprise. It also authorized the board 'to provide a system of organization to fix responsibility and promote efficiency'. The board decides upon major TVA programme, organization and administrative relationships. Responsibility for conducting TVA programmes, applying policies and methods, and performing services is delegated to the major organizational units.

There is a general manager appointed by the board to handle day-to-day operations of the agency. TVA programme activities are handled by two major groups based on budget procurement:

- power generation and transmission is a self-financing operation deriving

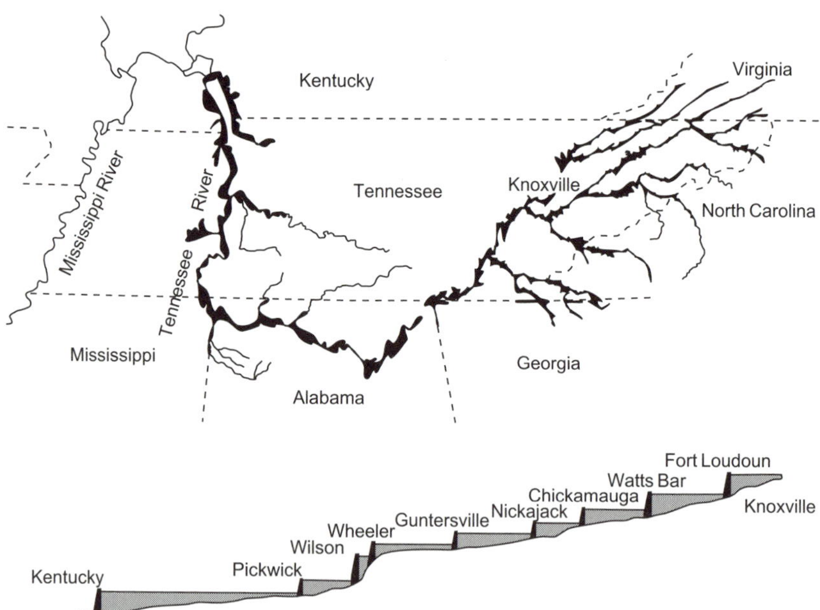

Figure 11.1 Map of Tennessee River watershed and diagrammatic profile of main stem dams and reservoirs (Source: Tennessee Valley Authority 1990).

funds from the sale of electric power to 160 distribution systems and selected industrial and government customers;
- the resource group programmes are funded primarily through congressional appropriations.

TVA serves part of a seven-state region in the south-east of the USA, which includes all of Tennessee and portions of Georgia, Alabama, Mississippi, Kentucky, Virginia and North Carolina (Figure 11.1). The heart of this region consists of the rivers and streams that make up the Tennessee River drainage basin. There are 125 counties that make up the Tennessee River watershed. This area covers 106,000 km^2 with a population of 8.5 million people (1992 estimate).

11.3 Project purpose and operational priorities

At the time the TVA was established, development of the Tennessee River was limited to two types of single-purpose water management projects (Gartrell *et al.* 1981). The first of these was a series of low head dams and canals constructed solely for improvement of navigation, especially in the interest of commerce. The other type, consisting of only two projects, had a dual purpose: hydrogeneration, and navigation improvement. In the late 1930s the TVA introduced a third type, the multipurpose project with three basic purposes: flood control, navigation, and hydrogeneration. Flood control was given the highest priority for operation, followed by navigation, with hydrogeneration third but strongly promoted as long as it was consistent with the other priorities. These priorities were established with the TVA Act and remain the basis for operation today. In general, managing the reservoir system for flood control, navigation, and hydrogeneration is compatible with mosquito and aquatic vegetation control. However, other programme interests, such as recreation, fish and wildlife, water supply, residential and other types of shoreline development, and other types of customer requests, create ongoing conflicts in demands on the system (TVA 1990).

Because of these conflicting demands, it is becoming increasingly difficult to incorporate water level management measures for mosquito and aquatic vegetation control into reservoir system operations. Recent proposals, primarily from recreational interests, suggest holding lake levels higher and essentially stable for longer periods of time during the summer months, thus increasing recreational benefits. Ironically this action could actually negate some of the anticipated benefits by adding to mosquito and aquatic plant problems, which directly impact on recreational use.

There are now forty-eight dams in the Tennessee River basin operating as a part of TVA's integrated water control system. Thirty-five are TVA dams, thirty of which are classified as major dams: nine main river projects and twenty-one tributary projects. The other five are classified as minor dams (Gartrell *et al.* 1981). The reservoirs of principal interest for mosquito control are the nine main projects (Kentucky, Pickwick, Wilson, Wheeler, Guntersville, Nickajack,

Chickamauga, Watts Bar, Fort Loudoun), although several of the tributary projects may be locally troublesome.

Because of the ever increasing demands being made on the system and the potential for significant programmatic conflict, a reservoir operations co-ordinating committee meets each year to hear all requests, address issues, and set operating priorities. This team is comprised of representatives of all programme interests to ensure that all requests are addressed, and includes: flood control; power generation (including hydro, fossil fuelled, and nuclear fuelled); navigation; recreation; fish and wildlife; archaeological; water supply; water quality; land resources; reservoir resources re-evaluations; community and economic development; and mosquito and aquatic vegetation control. Customers outside the agency also make their requests known through appropriate programmatic channels.

11.4 Public health and recreation

As already noted, the main reason for creating the TVA in 1933 was to transform this region of economic and social depression into a model of regional development. One of the major contributing factors to the existing social depression of that time was malaria, which affected the health and well-being of a large segment of the valley population. Even before impoundment of reservoirs began in the area, conditions existed that were favourable to the major malaria carrier in the south-east, *Anopheles quadrimaculatus*. Depressions in the floodplain along the major streams of the region, such as limestone sinks and potholes, held water for sufficiently long periods of time following floods or heavy rains to provide larval habitat for mosquitoes.

When TVA began its work, surveys in the vicinity of the proposed Wheeler Reservoir revealed a malaria prevalence as high as 60 per cent in humans in some areas, based on positive blood smears (United States Public Health Service and TVA 1947). This disease not only produced mortality but was also very debilitating to inhabitants of the region, hampering their ability to work and earn a living. It was clear, therefore, that TVA had to establish a programme at the very beginning to control this severe health problem with its broad implications for the economic and social well-being of the people of the region.

Based on studies and recommendations by the US Public Health Service in the early 1900s, regulations had been adopted by various southern states governing the conditions under which water might be impounded. The purpose was to minimize the potential hazard to public health. These regulations generally specified that any person, corporation, or agency desiring to impound water or change the levels of existing impoundments must first obtain a permit from the respective State Board of Health. The regulations further specified certain vector control actions that were to be taken to ensure that disease-transmitting mosquitoes were controlled (United States Public Health Service and TVA 1947).

Foremost among these specifications were guidelines for reservoir basin preparation, water level regulation, shoreline drainage, and floatage and vegetation control. With these regulations already in place in the valley states, TVA was legally obligated from the very beginning to include mosquito control in its regional development plans. In support of these regulations and to ensure compliance, TVA prepared a series of vector control specifications for each of its planned reservoirs. These were equal to, or surpassed, state requirements. In this way, legislative requirements were important in ensuring that mosquito control would be incorporated into impoundment plans for all water resource projects. Legislation of this type may be the best way to ensure that priority is given to vector control in planning and developing water projects. It also helps to eliminate conflicts among competing interests.

Community public health ordinances may also be effective, if properly enforced, as a complement to a regional control programme. Local mosquito problems are often related to poor environmental sanitation, such as inadequate solid waste disposal and improper discharge of industrial and domestic wastes. These conditions favour development of mosquito larval habitats. TVA regional programmes in solid waste and water management provide technical assistance to communities and help to eliminate mosquito habitat.

There are safeguards against pollutant discharges into the Tennessee River system provided under Section 26a of the TVA Act. Permits for discharge can be obtained only after favourable review. This provides TVA with another opportunity to reduce the potential for mosquito breeding in the reservoir system.

In addition to legislative requirements for vector control based on the impounded water acts of the seven valley states, TVA developed its own requirements for vector control. Administrative codes were developed stating agency policy and delegating responsibility to the appropriate programme organization within TVA. These statements specifically stipulate that vector control will be an integral part of the planning and operation of all water resource development projects. In this way, it was clear that vector control had the obvious support of TVA at its highest management level (Brooks 1987).

Further evidence of TVA support for vector control is contained in its environmental impact statement published in 1974 for the agency's vector control programme. This documents TVA's commitment to conduct vector control operations and to do so in accordance with environmentally acceptable practices. This commitment was subject to review by other federal and state agencies and by the general public before it was accepted for publication as part of the environmental impact statement. A supplement to the 1974 environmental impact statement is currently in preparation and discusses changes and impacts since the original statement was accepted.

The fact that malaria had a high incidence in the Tennessee Valley region, coupled with the focus of regional development to harness the Tennessee River into a series of slackwater reservoirs, and the knowledge that *Anopheles quadrimaculatus* breeds in these types of situations, created concern that a problem of unprecedented proportions could evolve. Consequently, the TVA entered into a co-operative malaria eradication programme with the US Public Health Service, working through the seven state health departments. The programme had a three-pronged approach consisting of measures directed at controlling populations of the mosquito vector, measures designed to reduce man/vector contact, and measures directed at eliminating the aetiological agent of malaria from the human population in the south-eastern USA. This was accomplished by 1949; no indigenous cases have been reported since.

TVA's efforts were focused primarily on controlling populations of the mosquito vector and reducing man/vector contact. Depopulation of the land area within 1.6 km of the margin of the reservoir to be impounded and mosquito-proofing dwellings were the thrust of the activities designed to reduce man/vector contact. These efforts, although very successful in helping to eliminate malaria from the valley, are in conflict with the current philosophy of optimizing recreational usage of the reservoir system with its inherent trend of increasing water exposure activities. Water-based recreation is directly linked to the quality of the water and the surrounding environment, therefore a discussion of reservoir water quality and recreational activities, potential disease exposure pathways, and TVA's management practices to address these concerns is appropriate here.

11.4.1 Water quality

Overall, the Tennessee River is generally considered to be a clean river (TVA 1990). Point and non-point sources of pollution degrade water quality at several locations on mainstream reservoirs and tributary rivers and reservoirs. Tables 11.1 and 11.2 summarize the principal water quality concerns in TVA reservoirs and from Tennessee Valley watersheds, respectively.

There is no one pervasive water quality concern in TVA reservoirs, but a collection of concerns affecting various uses. As shown in the tables, more major water quality concerns have been found on navigation and western tributary reservoirs than on eastern tributary reservoirs, and these concerns are more often caused by non-point sources of pollution than point sources. The specific concerns include PCB contamination of fish in the navigation reservoirs in the eastern half of the valley; aquatic plants from Chickamauga through Kentucky; pollution from major population centres affecting Boone and Fort Loudoun; pulp and paper mills affecting Douglas and Chickamauga; past DDT production near Huntsville, Alabama, affecting Wheeler; past mining activities affecting Nolichucky and the Ocoee reservoirs; and other non-point sources affecting Normandy, South Holston, Cherokee, Douglas, Watts Bar, Chickamauga, Guntersville, and Kentucky (TVA 1990).

Non-point sources are the cause of most principal water quality concerns affecting Tennessee Valley water. Agricultural and mining activities cause siltation and bacterial contamination that affect aquatic life and recreational uses of streams in the region. Most of the other principal water quality concerns are caused by present and past point sources of pollution affecting specific reaches of rivers in the Holston, French Broad, and Little Tennessee River watersheds.

The land adjacent to rivers, streams, and lakes in the Tennessee Valley serves multiple purposes. It is critical to the life-cycle of fish and wildlife and to the plants and organisms on which they depend for food and cover. It is also important for residential and industrial development, agriculture, mining, recreation, tourism, and other uses. Soil erosion and run-off as a result of residential, industrial, and recreational development increase non-point source pollution. In addition, such land use activities conflict with the preservation of the shoreline for scenic and aquatic resources, wildlife, and wetlands to the degree that natural shoreline habitat is removed or significantly altered.

People from across the nation are attracted to mainstream and tributary reservoirs that support over 200 species of fish. Large populations of native species (such as largemouth bass, crappie, buffalo, and catfish) have developed in many reservoirs and support important sport and commercial fisheries.

The impoundment of the main Tennessee River and its tributaries dramatically altered the river-stream habitat and the resulting food chain. Changes in the species composition of the fish community and in the number of surviving species occurred rapidly. Those species whose annual migration cycles were interrupted by dams, and those species whose requirements for temperature, spawning habitat, or food were closely associated with riverine conditions, rapidly declined in numbers or were eliminated.

The TVA attempted to enhance fish spawning by providing stable pool levels in reservoirs for about a six-week period during the peak of the spring spawning season. However, throughout the rest of the year, current operating procedures for tributary reservoirs are detrimental to fish populations and angler use. The most productive region of a reservoir is the shoreline because of submerged vegetation for cover and organics, nutrients, and aquatic invertebrates (benthos) for food.

On both tributary and mainstream reservoirs, shoreline development operates to modify fish habitat and other environmental factors that shape the fish community. Removal of vegetation in and near the water as a result of agricultural, industrial, residential and urban development subjects the area to more non-point source pollution from run-off on nearby lands. Mining, deforestation, domestic and industrial effluents, erosion, agricultural practices, and urbanization have affected nearly all of the fish habitat in the Tennessee River watershed.

Wetlands are transitional ecosystems between terrestrial and aquatic communities where the land is saturated with water or covered by shallow water during at least part of the growing season. Wetlands can become established in areas where frequent flooding occurs, nutrients are abundant, water flows are of low velocity, and soils allow the development of hydrophytic vegetation. Many wetlands are productive systems, owing to the combination of abundant water and nutrients that allow wetlands to develop.

Wetlands are given special consideration under federal laws and regulations because of their importance to aquatic life and because they provide habitat for certain wildlife species. Wetlands have also been shown to be important for

Table 11.1 Principal water quality concerns in Tennessee Valley Authority reservoirs

Reservoir	*Location*	*Uses affected*				*Source*	
		Aquatic life	*Fish consumption*	*Recreation*	*Water supply*	*Point*	*Non-point*
Kentucky	Navigation and west tributaries	Dissolved oxygen		Aq. plants			×
Normandy	"	Dissolved oxygen			Taste, odour, Fe, Mn		×
Pickwick	"						
Wilson	"				Taste, odour		×
Wheeler	"		DDT	Aq. plants		×	×
Tims Ford	"						
Guntersville	"			Bacteria, aq. plants		×	×
Nickajack	"		PCBs	Aq. plants			
Chickamauga	"	Dissolved oxygen		Aq. plants, colour		×	×
Watts Bar	"	Dissolved oxygen	PCBs			×	×
Melton Hill	"		PCBs	Aq. plants			×
Fort Loudoun	"	Dissolved oxygen	PCBs	Bacteria		×	×
Tellico	"		PCBs				
Norris	East tributaries						
Cherokee	"	Dissolved oxygen				×	×
Fort Patrick	"						
Boone	"	Dissolved oxygen	Metals, toxics	Bacteria		×	×
South Holston	"			Bacteria			×
Wilbur	"						
Watauga	"						

Table 11.1 continued

Reservoir	*Location*	*Uses affected*				*Source*	
		Aquatic life	*Fish consumption*	*Recreation*	*Water supply*	*Point*	*Non-point*
Douglas	"	Dissolved oxygen			Colour	×	×
Nolichucky	"	Siltation		Siltation	Siltation		×
Fontana	"						
Ocoee 1–3	"	Metals, siltation		Siltation		×	×
Blue Ridge	"						
Appalachia	"						
Hiwassee	"						
Nottely	"						
Chatuge	"						

Source: Tennessee River and Reservoir System Operation and Planning Review (TVA 1990).

Table. 11.2 Water quality concerns from Tennessee Valley watersheds

	Uses affected[a]					Source	
Watersheds	*Aquatic life*	*Fish consumption*	*Recreation*	*Water supply*	*Future development*	*Point*	*Non-point*
Tennessee							
Pickwick	Toxics		Bacteria				×
Watts Barr	Siltation		Bacteria				×
Duck	Siltation		Bacteria				×
Sequatchie	Siltation		Bacteria				×
Clinch/Powell	Siltation		Bacteria				×
Holston							
North Fork		Mercury	Bacteria			×	×
South Fork	Siltation	Metals, toxics	Bacteria		Limited assimilative capacity	×	×
French Broad	Siltation	Dioxin	Bacteria			×	×
Little Tennessee	Siltation	PCBs				×	×
Hiwassee	Metals, siltation		Bacteria	Bacteria		×	×

[a] Uses are affected by the problem noted on at least one stream in the watershed.

Source: Tennessee River and Reservoir System Operation and Planning Review (TVA 1990).

erosion control, flood and storm damage attenuation, water quality improvement, and ground water recharge.

The TVA is acquiring data on wetlands from aerial surveys of aquatic plants and as part of its reservoir lands planning process. However, detailed information about wetlands in the Tennessee Valley is lacking at this time. Based on the field experience of TVA staff, it is estimated that over 90 per cent of the wetlands on TVA reservoirs are located on mainstream reservoirs. Tributary reservoirs have few wetlands because of the steeper slope of their shorelines, and the steeper drawdown for flood control. The topography around mainstream lakes is flatter, lending itself to the establishment of wetlands, and there is much less drawdown from summer to winter on mainstream lakes. In addition, there is about three times as much shoreline on mainstream lakes as there is on tributary lakes.

The groups of wildlife most closely associated with streams and reservoirs are: waterfowl (ducks, geese and swans); waterbirds (loons, herons and cormorants); and wetland furbearers (muskrats, beavers, mink, otter and raccoons). The animals within these groups are dependent to varying degrees upon streams, reservoirs, and the lands bordering them for feeding, nesting sites and shelter.

11.4.2 Recreation

Reservoirs in the valley region are used for a variety of water-oriented recreational activities. Swimming, fishing, water-skiing, and boating are enjoyed on the lake, and camping, hiking, picnicking, sightseeing, nature-watching, and fishing are enjoyed from the shoreline. A variety of factors affect the type and amount of recreational activities that occur on any given lake. Among the most important are lake surface area (including miles of shoreline), annual and summer drawdown, access, location (including proximity to population centres), water quality, and reservoir aesthetics. Another characteristic affecting resource use in the reservoir system is the presence and relative abundance of mosquitoes and related medically important arthropod pests.

Consequently, in developing a recreational resource development policy, consideration of impacts of this type and how they could be mitigated were addressed in defining permissible recreation activities, locations, etc.

From its beginning, the TVA has worked to encourage development of a wide variety of outdoor recreational facilities and opportunities in the Tennessee Valley, particularly on TVA lakes and shorelines. To secure the involvement of the people most affected, TVA structured its recreation policy so that its activities would stimulate, support, and complement the actions of concerned agencies and individuals.

In summary, the policy identifies the recreational resources available throughout the valley (especially those associated with TVA lakes), encourages development of other public agencies and private investors, provides technical assistance where needed to achieve this development, and provides basic facilities where needed to assure safe access to the lake and to protect the shoreline from misuse (TVA 1987).

TVA has set aside large tracts of reservoir shoreline for wildlife management and hunting areas, for wildlife refuges, and for duck and geese feeding areas to be managed by the states' game and fish agencies and the US Fish and Wildlife Service. Also, approximately 50,000 ha of reservoir lands have been transferred to the National Park Service and the US Forest Service, which administer them largely as natural forest lands.

Since 1969, the TVA has been providing basic recreational improvements such as picnic facilities, boat-launching ramps, access roads, and sanitary facilities at areas along its reservoir shorelines where public use indicates a need for such facilities. By 1978 use of these shoreline facilities had grown to such an extent that it became evident to the TVA Board of Directors that a new policy for facility management was needed. During 1979, TVA began implementing the new policy. To ensure that unique local conditions were taken into account, community meetings were held across the Tennessee Valley to invite comment by valley leaders and citizens. Based on these comments, specific management plans were developed for management of TVA facilities on reservoir properties (TVA 1987).

The new policy basically maintained TVA's long-standing recreational goals: to provide a quality outdoor experience for citizens of the valley and for visitors to the area; to encourage state and local government agencies to develop parks and other recreational facilities wherever feasible; to assist in the growth and development of quality private recreational facilities wherever feasible; and to assist in the growth and development of quality private recreational opportunities in the valley. In management of its own facilities, the TVA designated certain areas for specific uses (day use, overnight camping, informal use including a combination of camping and day use, boat launching). The TVA began employing onsite, resident caretakers for key areas, and in return a modest fee was charged for overnight camping at developed campgrounds.

Through planning and technical assistance, TVA furnished guidance to all types of parks and recreation programmes as well as to development of lake-oriented facilities. It also furnishes information about recreational use and development of the region's resources for TVA and other planners to use in analysing and evaluating recreational opportunities and needs.

One of the most effective ways TVA supports its overall recreational commitment has been to make suitable portions of its shoreline lands available to others for development. Land and land rights (92,126 ha) have been transferred or conveyed for a nominal consideration to federal, state, and local government agencies for the development of public parks and access areas (Table 11.3). Lands have been leased, licensed, and sold to quasipublic groups and organizations for group camps. In addition to these activities on reservoir lands, TVA has provided a wide range of technical assistance to communities, counties, municipalities, and state agencies to help them improve their own recreational programmes. In March 1987, according to Division of Land and Economic Resources records, the TVA watershed had 55 group camps or clubs, 298 commercial recreation areas and approximately 37,000 boats or houseboats moored on the lakes.

In addition to the more reservoir-based recreational activities, canoeing, river fishing, hiking, and biking have grown tremendously in popularity, and TVA is working with various groups to promote protection of streams while providing for their use. During 1978, TVA began acquisition of some 250 stream access sites on 40 scenic streams throughout the valley. TVA is providing assistance to commercial river outfitters and is deeply involved in public information related to stream recreation (TVA 1987).

TVA is also concentrating the development of hiking and biking trails on agency lands near population centres and developed recreation clusters such as 'land between the lakes' (LBL). A significant addition to the recreational opportunities associated with TVA impoundments, LBL was established as a centre for outdoor recreation and environmental education and resource stewardship, managed by TVA as a national demonstration. Methods and principles for demonstration, development, and operation of the area are being tried and tested to help establish criteria for managing multiple use facilities and resources throughout the USA.

LBL is a 68,798 ha wooded peninsula in western Kentucky and Tennessee. It is between Kentucky Lake, built by TVA, and Lake Barkley, a US Corps of Engineers project. The area is roughly 13 km wide, 64 km long, 85 per cent wooded, and has 480 km of shoreline.

LBL receives more than two million visits annually from a wide variety of user groups, ranging from hikers and campers to hunters and fishermen. Recreational facilities include three family campgrounds that provide 900 tent and trailer sites; 27 informal day-use areas that provide additional camping space and day-use facilities; and Brandon Spring Group Camp with housing and dining facilities.

LBL is a continuing development and demonstration project. There are special areas provided for off-road vehicles, campers with horses, and field archers; and school groups can come for residential programmes in environmental education at the Youth Station, an outdoor school located in the Environmental Education Area, 2,023 ha along the shore of Lake Barkley (TVA 1987).

Visitation to mainstream reservoirs is about three times higher than visitation to tributary reservoirs (Table 11.3). Investments in recreational facilities and homes are also about three times higher on mainstream than tributary reservoirs. Mainstream reservoirs (including tributaries in the western half of the valley and reservoirs with commercial navigation channels) have a total of 190,206 ha of surface area during the summer recreation season. They are drawn down an average of 2 m annually, and only 0.3 m during the summer. By comparison, tributary reservoirs in the eastern half of the valley have about 55,000 ha of surface area during the summer recreation season, less than 30 per cent of that in mainstream reservoirs. Moreover, the size of tributary area lakes can vary considerably because their drawdown is much steeper: an average of 9.75 m annually and 4.57 m during the summer.

Until relatively recently, fees for using TVA public use facilities were minimal or non-existent since much of their development was funded through taxation by governmental appropriation. A recent reduction in annual governmental

Table 11.3 Recreational visits and facilities on Tennessee Valley Authority reservoirs

				Number of facilities						
Reservoir	*Location*	*Annual visitor days*	*Rec. facil. invest.*	*State parks*	*City county parks*	*Camps, clubs*	*Comm. rec. areas*	*Public access facil.*	*Wildlife mgt areas*	*Other nat. areas*
Kentucky	Navigation and west tributaries	4.0	180.0	4	6	10	92	38	3	31
Normandy	"	N/A	N/A					6		
Pickwick	"	1.9	62.0	2	5	2	11	14	4	8
Wilson	"	1.2	43.0		1	2	8	4		
Wheeler	"	1.0	42.0	1	6	2	9	12	3	1
Tims Ford	"	0.2	8.3	1	2	1	3	7		
Guntersville	"	2.2	130.0	2	11	7	28	30	4	3
Nickajack	"	0.2	6.8		2		6	7		2
Chickamauga	"	2.0	140.0	2	6	9	15	82	2	
Watts Bar	"	1.0	71.0		7	6	31	77	2	4
Melton Hill	"	0.2	9.0		6	2	6	5		
Fort Loudoun	"	0.6	70.0		12	2	13	37		
Tellico	"	N/A	N/A					12		
TOTAL		14.5	762.1	12	64	43	222	331	18	49

Table 11.3 continued

Reservoir	Location	Annual visitor days[a]	Rec. facil. invest.[b]	Number of facilities: State parks	City county parks	Camps, clubs	Comm. rec. areas	Public access facil.	Wildlife mgt areas	Other nat. areas
Norris	East tributaries	1.40	77.00	3	6	5	28	64	2	11
Cherokee	"	0.60	26.00	1	6		15	29	1	0
Fort Pat	"	0.50	12.00	1	1			3		
Boone	"	0.30	34.00		1	2	8	14		
South Holston	"	0.05	13.00		2	1	7	3	1	
Wilbur	"	0.01	0.01							
Watauga	"	0.30	9.70				6	4		
Douglas	"	0.50	18.00		2		10	9	2	
Nolichucky	"	0.01	0.50		1					2
Fontana	"	0.20	8.70		2					
Ocoee No. 1	"	0.10	3.30			2	8	1		
Ocoee No. 2	"					5	1	4	1	
Ocoee No. 3	"					1				
Blue Ridge	"	0.10	5.30		1	2	1			
Appalachia	"	0.10								
Hiwassee	"	0.10	10.00		2		3			
Nottely	"	0.10	4.80		1		7	2		
Chatuge	"	0.50	17.00		3	1	15	3		
TOTAL		4.87	239.31	5	28	19	109	136	7	13

Source: Tennessee River and Reservoir System Operation and Planning Review (TVA 1990).

[a] Visitor days in units of 100,000.

[b] Investment in units of US$100,000.

appropriations for TVA, resulting from a political desire to reduce the size of government and hence the taxpayer burden, and to require the benefiting public to pay for services received, has forced the agency to look to this group for more monetary support. The process has begun, and eventually most recreational programmes will be self-supporting or receive only minimal governmental appropriation support. In 1994, revenues from the rental of twenty-five picnic pavilions was US$18,988 and from thirty-five fee campgrounds, US$517,203. These figures are exclusive of LBL; revenues from LBL for 1995 were US$2,770,000.

Since recreation has become a major focus in the use of the reservoir system and is actively promoted by TVA, the agency strives to maintain the reservoir environment in a condition that is suitable for human use and development as well as suitable for animal habitation. To ensure that these conditions exist and that mosquitoes are not a major impediment to use, TVA continually monitors these conditions and applies remedial control measures as necessary.

Mosquito population monitoring involves both the permanent pool species as well as the floodwater groups. During the breeding season for the permanent pool group, generally considered to be May to September, although this will vary depending on water levels on individual reservoirs, adult and larval populations are monitored weekly at about 150 stations on the nine mainstream and several tributary reservoirs. Mosquito 'inspectors' initially examine index stations for adults. If adults are detected, a more intensive search for larvae in the pool is conducted within the immediate vicinity of the index station. The data are reported weekly to the programme headquarters in Muscle Shoals. When any given inspection shows that the larval index exceeds the treatment threshold of 0.25 larvae per sample, this information is relayed on the same day so that control applications can be scheduled. If the adult index exceeds fifteen female *Anopheles quadrimaculatus*, this information is also relayed immediately for planning control measures.

These two indices together ensure that individuals engaged in outdoor activities, especially recreational, have a quality experience, relatively free of the annoyance and health consequences of exposure to blood-sucking arthropods. The larval index was derived from an assessment of data over a period of years that demonstrated that complaints did not occur and adulticide treatments were not required when the value remained below 0.25 larvae per dipper sample. The adult index is based on the level of annoyance that could be tolerated by individuals engaged in outdoor activities. The annoyance level was related to the number of adult female *Anopheles quadrimaculatus* mosquitoes occurring in diurnal resting stations, and what that number represented in terms of biting mosquitoes during their feeding period.

Floodwater mosquito populations of the genera *Aedes* and *Psorophora* are monitored annually by determining the number of eggs in soil samples collected from known areas that produce floodwater mosquitoes. The number of eggs per species is recorded with data relating to the elevation in metres above mean sea level at which they were collected (Figure 11.2). These data can then be used to

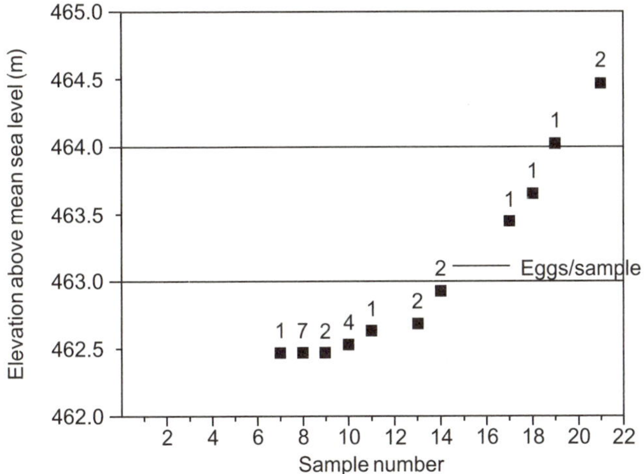

Figure 11.2 Profile of floodwater mosquito egg deposition in the flood zone of a typical reservoir embayment.

delineate the extent of pre-flood applications of larvicide or to prepare for post-hatch larvicide as lake levels rise and approach egg bed elevations. Floodwater mosquitoes, although of only minor health concern to humans, are particularly detrimental to recreationists because of their aggressive nature, their propensity to feed during daylight hours, and the fact that they emerge periodically in vast numbers.

In addition to monitoring mosquito population levels, the TVA, in concert with state regulatory agencies, examines and investigates other issues that may evolve as lake-based recreation continues to increase. The TVA has no enforcement authority and must depend on state responses to its findings. The TVA monitors water quality for priority pollutants and coliform bacteria at fixed monitoring stations and reports its findings to the respective state regulators. These stations include monitoring at some of the major public campground sites to determine if domestic wastes generated there are ending up in the reservoir. This is unlikely since the majority of established public use areas are equipped with chemical toilets or septic tanks, which are drained periodically. Dump facilities are also provided for toilet-equipped recreational vehicles. State regulations exist that require boats with overnight-type accommodation to have holding tanks for domestic waste, which must be properly disposed of in port. The TVA, again in conjunction with state regulators, polices the waters and public launch facilities with concern for water safety and related public health issues. Major water-based activities such as fishing competitions, tournaments, organized pleasure cruises (flotillas), etc. are regulated under appropriate state laws and are monitored by the states with support from the TVA.

In conjunction with the other services provided by the TVA to promote recreational development, the agency marks the secondary navigation channels with appropriate buoys to help safeguard the public from underwater obstructions. However, many of the services currently provided with federal tax dollars may be significantly reduced or eliminated, and the work and associated costs transferred to state agencies or the benefiting public. For example, although respective valley states are responsible for managing the water system fishery, including the stocking of fish, TVA has in the past provided monetary support for this work. Future monetary support will be solicited from sportsmen's groups, e.g. 'Trout Unlimited', or other benefiting user groups. This trend of requiring the user to pay will become more pronounced in future years as government appropriations decline in accordance with political mandates to reduce the size of federal government and budgets. Theoretically, personal tax liabilities should decrease and consumers should have more money to spend on personalized recreation. Consequently, the private sector will become more involved in reservoir-based recreation and TVA, like other federal agencies, will retain only those properties that must be held in public trust to ensure wise stewardship.

With its mandate of environmental stewardship and its commitment to be a leader in environmental quality, the TVA generally goes beyond the minimum in complying with federal and state regulations. In fact in many situations TVA has established standards that other agencies are trying to match. However, this environmental ethic is not necessarily compatible in all respects with vector

Figure 11.3 Beaver dam constructed across a natural creek drain creating a mosquito breeding pond.

control and sometimes produces direct conflicts of purpose. The Clean Water Act of 1972 and all of its amendments create some obvious concerns for vector control programmes, as does the recent federal policy on wetlands which specifies no net loss. These two together create difficulty for programmes that rely heavily on environmental management measures such as water level manipulation and its necessary companion, drainage ditch maintenance. The dewatering action coupled with ditch maintenance is being challenged as having a negative impact on wetlands even though these activities may have been in place for fifty years or more and are simply maintaining the status quo. Moreover, the presence of beavers, *Castor canadensis*, in drainage systems adds to the problem by constantly creating ponds.

Beavers have became particularly bothersome in the Tennessee Valley region because of their dam building activities (Figure 11.3). They have reached nuisance proportions in many areas of the reservoir system because they have no effective natural predators, and no market currently exists for the fur or meat. In the course of one to two days, they can create ponds by blocking flow in constructed ditches or in natural drains, thereby creating very productive additional mosquito habitat. If the created pond goes undetected and is not removed within a year, it is considered to be wetland and is protected from removal or destruction unless the required permits are obtained. TVA's vector control programme monitors this extensive network of constructed ditches and natural drains, and when beavers are detected in the critical areas they are rapidly removed by trapping and the dams breached to allow normal flow. Although beavers can provide limited additional opportunities (fishing and hunting), the loss of valuable timber resources and the adverse impacts to other recreationists probably results in a net loss in the value of the area.

Ironically, policies that have been designed to protect wetlands and improve environmental quality can have detrimental impacts by changing the habitat from a flowing-water to a static-water environment, with the concurrent loss of specific types of aquatic life and the creation of mosquito habitat. Consequently, the loss of the use of environmental management measures for mosquito control results in an increase in the use of insecticides. Although the use of biocontrol agents is emphasized, the number available for use is very limited, and insect resistance has already evolved in a number of species (Tabashnik 1994). Again it is apparent that regulations and policies allegedly designed to protect environmental integrity promote some species at the expense of others; often this is dictated by the mission or interest of specific agencies and their programmes, e.g. waterfowl management, fishery enhancement, endangered species protection, to name just a few.

The Endangered Species Act of 1973, with its well meaning intention of protecting and preserving threatened or endangered species, limits its actions to 'listed species' only. Consequently, dominant community members of some habitats may be sacrificed for the sake of a listed species, causing shifts in aquatic community structure. The act has also caused a review of all pesticides used in mosquito/vector control to determine which compounds could still be generally used. Currently, only one compound, *Bacillus thuringiensis israelensis* (*Bti*), remains listed for this purpose. Therefore, all others must receive extensive review and permitting before they may be used in areas where threatened or endangered (T&E) species are thought to occur, regardless of any known specific adverse impacts on the listed species. The curtailment of the use of insecticides in mosquito control can also jeopardize listed species by exposing them to mosquito-borne pathogens, e.g. the mosquito-transmitted eastern equine encephalitis virus is a serious threat to the remaining population of whooping cranes. Other mosquito-transmitted zoonoses exist that affect not only T&E species but unlisted animals as well.

The 1972 amendments to the Federal Insecticide Fungicide and Rodenticide Act (FIFRA) classified pesticides into two broad categories: general use and restricted. It further required that all individuals applying pesticides classified as restricted must be certified. Consequently this level of regulation should be adequate to ensure that pesticides are properly applied in accordance with regulations that protect the environment, making it appear that some redundancy exists between the regulations. Nevertheless, the regulations do exist and compliance is required. So even though, on the one hand, control options are available that would seem to permit a wide range of uses of the water resource, including recreation, on the other hand other regulatory constraints may contradict their use. To counter these constraints and ensure that adequate control options are available, recreation and the mosquito control activities that support it must be given a high priority in the planning and subsequent management of water resource development projects. Mosquito control legislation may be the best way to ensure that this important work receives the attention it deserves.

References

Brooks, R.H. (1987) Tennessee Valley Authority: Institutional Arrangements for Health and Environmental Protection in Water Resource Management. In: *Selected Working Papers of the III, IV, V, and VI Meeting of the WHO/FAO/UNEP Panel of Experts on Environmental Management for Vector Control.* WHO/VBC 87.3, World Health Organization, Geneva, pp. 45–51.

Gartrell, F.E., Cooney, J.C., Chambers, G.P. and Brooks, R.H. (1981) TVA Mosquito Control 1934–1980 – Experience and Current Program Trends and Developments. *Mosquito News*, **41**, 302–322.

Tabashnik, B.E. (1994) Evolution of Resistance to Bacillus. *Annual Review of Entomology*, **39**, 47–79.

TVA (1987) *TVA Handbook.* Tennessee Valley Authority, Knoxville.

Tennessee Valley Authority (1990) *Tennessee River and Reservoir System Operation and Planning Review. Final Environmental Impact Statement.* TVA, Knoxville.

United States Public Health Service and Tennessee Valley Authority (1947) *Malaria Control on Impounded Water.* US Government Printing Office, Washington, DC.

Chapter 12

The Tennessee Valley Authority programme to control mosquitoes in fresh water impoundments and wetlands

Edward L. Snoddy (deceased) and Joseph C. Cooney

12.1 Introduction

Although from the beginning the Tennessee Valley Authority (TVA) placed principal reliance for mosquito control upon biological or naturalistic control measures, extensive use was made of supplementary measures in the form of chemical larviciding (TVA and the United States Public Health Service 1947). Between 1933 and 1943, developments in larviciding techniques consisted principally of improvements in equipment for hand and boat application of kerosene and black oil larvicidal mixtures and aerial applications of Paris green dust. As DDT and other more effective insecticides became available, the TVA began testing the more promising ones and developing satisfactory equipment for their application. Aeroplane application of DDT was first carried out by the TVA in co-operation with the US Department of Agriculture in 1943. Because of the demonstrable effectiveness of DDT at low rates of application (0.05–0.10 kg/ha) and the resultant economics of aerial distribution, DDT was adopted for TVA use in mosquito control operations. The use of Paris green dust for aerial applications was completely phased out by 1946 and boat applications of kerosene and black oil mixtures were discontinued in 1949. Later, spray equipment was devised to replace the aerosol generating equipment. In the early 1950s, the use of fixed-wing aircraft equipped to produce an aerosol or mist type spray was evaluated, but their use was discontinued shortly after it began.

The overall objective of TVA's present mosquito control programme as defined in the latest revision of operational procedures is to safeguard the health and well-being of the valley population affected by the production of mosquitoes on TVA lands and waters. The goal is to suppress mosquito populations in designated critical areas below a level at which disease transmission and human annoyance occur. Control operations are not routinely applied reservoir-wide but adhere to the 'critical area' concept:

- sites where aggregations of people occur within mosquito flight range of a reservoir;

- areas from which complaints originate;
- areas in which control operations were required the preceding year;
- those areas which during the preceding year developed mosquito breeding potential. These may include residential areas, recreational sites, commercial and industrial locations, educational facilities, and other locations where people tend to aggregate on or close to the reservoirs.

The TVA control programme is directed primarily towards the control of permanent pool mosquito species such as *Anopheles quadrimaculatus* and *Anopheles punctipennis*, although specific targets also include floodwater species such as *Aedes vexans*, *Aedes sticticus*, *Aedes trivittatus*, *Psorophora ferox*, *Psorophora cyanescens*, and *Psorophora mathesoni*.

In addition to the aforementioned species, problems may also occur in limited areas associated with breeding in organically contaminated pools, e.g. constructed wetlands developed to process domestic wastes. These areas may be limited in size but are potentially of great concern because of the species that breed under these conditions and the large number that can be generated. Species such as *Culex quinquefasciatus*, *Culex salinarius*, *Culex tarsalis* and *Aedes albopictus* are vectors of encephalitis viruses, dengue, and dog heartworm.

Many methodologies have been developed or improved through the application of research conducted by the TVA staff and others (Gartrell *et al.* 1981; Hammer and Kadlec 1983). Many of the older programme methodologies and techniques are described in *Malaria Control on Impounded Water*, (TVA and the US Public Health Service 1947). New ideas, methods, materials, and equipment are incorporated into TVA control operations as they are proved safe and effective through appropriate laboratory and field tests; these developments are reported extensively in the literature.

Although many of these techniques are still effective (Snoddy and Cooney 1989), conflicts with other programme interests preclude their use in many situations. Antipesticide sentiment has drastically curtailed the use of even the most inoffensive compounds. Consequently, TVA has been redirecting its research emphasis in recent years from studies on alternative pesticidal control techniques to evaluations of biological agents, focusing either directly on the mosquito target or its aquatic habitat.

12.2 Control of permanent pool mosquito species

Environmental management provides the basis for an integrated programme to control permanent pool and other mosquito species. The methodologies used are as follows:

- pre-impoundment preparation of reservoir basins
- shoreline alterations

- marginal drainage maintenance
- operation of dewatering projects
- mechanical control of shoreline emergent vegetation
- biological control of submerged vegetation
- water level management.

Despite continual refinement and efficiency in the use of these methods, applications of adulticides or larvicides are required to supplement these techniques.

12.2.1 Pre-impoundment – reservoir basin preparation

Prior to impoundment, all trees, bushes and other vegetation that would break the water surface at maximum drawdown level (and thus provide mosquito larvae with habitat) were removed. Initial clearing specifications were developed intersectorally so that the requirements of all programmes were met, including power, navigation, flood control, recreation and vector control. Clearing of the reservoir prior to impoundment was accomplished by hand or with machinery, and consisted of removing all types of woody vegetation that penetrated the water surface at minimum pool levels. If allowed to remain, this vegetative material can create a significant amount of mosquito habitat after the reservoir is filled. Despite the reservoir basin clearing operation, other vegetation may colonize the reservoir after it is filled and still create mosquito habitat, thereby requiring management. Consequently, removing water from the weeds or the weeds from the water is axiomatic when controlling permanent pool type mosquitoes such as *Anopheles quadrimaculatus*, *Anopheles punctipennis* and *Culex erraticus*.

The more recent impoundment plans contain requirements from other programme interests to allow timber to remain in situations where it would extend above the proposed water surface. The objectives were to enhance fish and waterfowl populations and to provide fish attractors that would benefit anglers. After careful study, it was determined that standing timber would not create mosquito habitat as long as water movement occurred, thus preventing the accumulation of floatage. Therefore, exceptions to the very stringent reservoir basin preparation guidelines were permitted and will be considered in future requests, although it is highly unlikely that TVA will construct additional reservoirs.

These granted exceptions are also consistent with a dramatic change in TVA's mosquito control philosophy that occurred in 1973. Since malaria transmission no longer occurred, the basic approach changed from an attitude of total anopheline control on all reservoirs to a strategy of selective control in 'critical areas' of various reservoirs. Continual suppression of anopheline populations lessens the potential for transmission of reintroduced malaria, and also reduces the annoyance level and other non-pathogenic conditions to humans and animals associated with mosquito bites.

A second type of mosquito that may be associated with TVA impoundments is the floodwater or temporary pool group such as *Aedes vexans*, *Psorophora ferox* and *Psorophora mathesoni*. These species have little vector importance to man although they transmit heartworm to canines and may be involved in zoonotic cycles of disease. They are, however, a severe annoyance to man and animals living in the vicinity of certain reservoirs when high flows and flood control operations are occurring. This situation usually occurs only once a year, usually during spring or early summer. Reservoir basin clearing or vegetation management has little impact on floodwater mosquito production; however, these operations could make follow-up control measures such as larviciding more efficient and effective.

12.2.2 Shoreline alterations

Shoreline alterations such as deepening or filling, or a combination of these, have been used successfully to reduce mosquito breeding habitat prior to impoundment, and in several cases after impoundment had occurred. These activities may consist of filling and levelling depressions in the fluctuation zone so that they will not retain water during drawdown periods, a deepening of the waterways, and creation of steeper shoreline margins by excavating original basins to create new banks above the original pool levels to create a new bank. This action reduces the littoral zone by decreasing the area in which aquatic plants will grow, subsequently reducing mosquito larval habitat.

12.2.3 Marginal drainage maintenance

The establishment of an effective drainage system is tantamount to having an effective water level management operation for mosquito control. Water recedes rapidly and completely from marginal reservoir areas during the drawdown period of planned fluctuations or flood operations. In most TVA reservoirs, a series of drainage ditches was established prior to initial impoundment to connect depressions in the normal fluctuation zone of the reservoir so that the water in these areas rises and falls with the reservoir. In some areas, ditches through the flood zone ensure rapid and complete recession of empty low depressions that may provide larval habitat.

Early ditches were constructed by hand or with explosives and earth-moving machinery such as backhoes and trackhoes. These ditches are routinely inspected and the functioning level monitored so that routine maintenance can be performed. When ditch maintenance is required, it is currently performed using a backhoe or trackhoe, or whenever possible a rotary ditcher (Figure 12.1). The rotary ditcher has rapidly become the equipment of choice because of its low cost per linear metre of excavation, its efficiency of operation, and its environmental acceptability. The ditcher leaves no unsightly spoil piles and broadcasts the excavated material evenly over an area of about 16 m from the ditch. The discharge is so fine and evenly spread that from a distance of about 16–25 m it is almost undetectable. Explosives are no longer used to excavate drainage ditches

Figure 12.1 Rotary ditcher.

because many of the ditches are no longer at a sufficient distance from structures to assure that no concussion damage will occur. Human development has been rapid around TVA reservoirs.

Recent initiatives in wetlands preservation and the issuance of new regulations regarding the drainage of wetlands are beginning to adversely impact TVA's ability to conduct this work. 'Grandfathering' policies allow TVA to continue to maintain the ditches that still function. However, any new construction requires extensive review and the acquisition of a permit from the United States Army Corps of Engineers (USACOE) in accordance with section 404 of the Clean Water Act. Consequently, it is imperative that ditches be closely monitored to ensure their functionality. Otherwise, for non-functional ditches, such reclamation would be considered new construction and subject to exhaustive environmental review.

12.2.4 Operation of dewatering projects

The construction of dewatering units was a novel way to control mosquitoes in a number of troublesome areas by permitting the water level in these projects to fluctuate independently of the main reservoirs. For areas of the reservoir basin that were too wet to allow adequate clearing and contained the potential to convert to breeding habitat after impoundment, dikes were established to separate these areas from the main reservoir. Dewatering of these areas was accomplished by pumping (Figure 12.2), particularly in late spring as the main reservoir rose during the mosquito breeding season. In some projects, dewatering facilitated the planting of row crops, which were partially harvested, leaving some as food for migratory waterfowl. As the main reservoir began its seasonal recession and mosquito breeding ceased, pump gates were reopened to allow fluctuation with seasonal lake levels.

This process was an effective control technique over relatively large areas when the philosophy of mosquito control was elimination of all anopheline larval habitats. However, with the changed philosophy to the critical area concept, this type of operation became economically unacceptable, especially in the light of the availability of newer and cheaper insecticides and dissemination systems. Consequently, in 1976, the operation of all dewatering units was transferred from the vector control programme to TVA's waterfowl management programmes. Although the units are now managed and paid for with waterfowl management

Figure 12.2 Dewatering project.

appropriations, some benefits still accrue to mosquito control if the units are operated in a dewatering mode. Operating schemes can be developed that benefit several programme interests (fish, wildlife, mosquito control) and the costs can be shared accordingly. As restrictions on the use of insecticides become more stringent, and vector and pathogen resistance become more widespread, development and operation of dewatering units for vector control will gain acceptance again, especially when the benefits to recreation are also realized.

12.2.5 Mechanical control of emergent vegetation

Mechanical control of marginal vegetation (tractor drawn rotary mowers) is accomplished every two or three years to remove the growth of vegetation in the drawdown zone that could provide breeding habitat during the next season. The two major species of woody vegetation targeted are buttonbush and black willow. This activity not only eliminates larval habitat but also removes impediments to larvicidal applications.

During each autumn, selective mowing (Figure 12.3) is a relatively minor activity involving no more than 80–120 ha annually. After the reservoirs fill during the following summer, plants remain in narrow bands and clumps along creek channels (Figure 12.4), drainage ditches, and as scattered individual plants in the deeper portions of embayments, to serve as identifiable food and cover for fish. Woody vegetation remaining under these conditions is sufficiently scattered to allow water movement, thereby minimizing mosquito habitat.

Figure 12.3 Selective mowing.

Figure 12.4 Selective mowing after inundation.

12.2.6 Biological control of submerged vegetation

As mosquito larvae are often found in marginal emergent vegetation, littoral zones are limited, where possible, to a narrow band of emergent plant species. Water level management worked exceptionally well for many years, especially on those reservoirs experiencing cyclical fluctuations in water levels. However, as reservoirs aged and native submersed aquatic species began to colonize (e.g. *Potamogeton* and *Najas*) and some of the erect stem shoreline emergent species were replaced by mat-forming emergent species (e.g. *Alternanthera*, *Polygonum*), mosquito populations of *Anopheles quadrimaculatus* and *Culex erraticus* began to increase.

The introduction of the exotic submersed aquatic plant Eurasian water milfoil (*Myriophyllum spicatum*) into the TVA reservoirs in the 1950s resulted in extensive habitat for the production of the permanent pool mosquito *Anopheles quadrimaculatus*. By 1968, milfoil had colonized over 10,000 ha on TVA reservoirs. After reaching the water surface or 'topping out', water milfoil forms interstitial pools that protect mosquito larvae from predation, and also provides a food source by harbouring algae and other organic debris (Figure 12.5). By 1980 at places such as the Chickamauga reservoir, east Tennessee, mosquito populations had essentially exploded (Figure 12.6)!

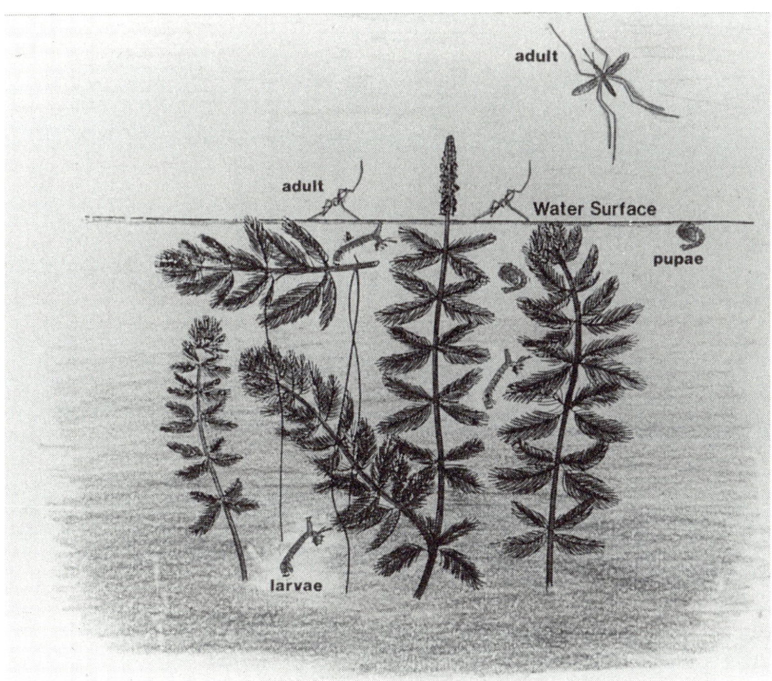

Figure 12.5 Larval habitat created by milfoil.

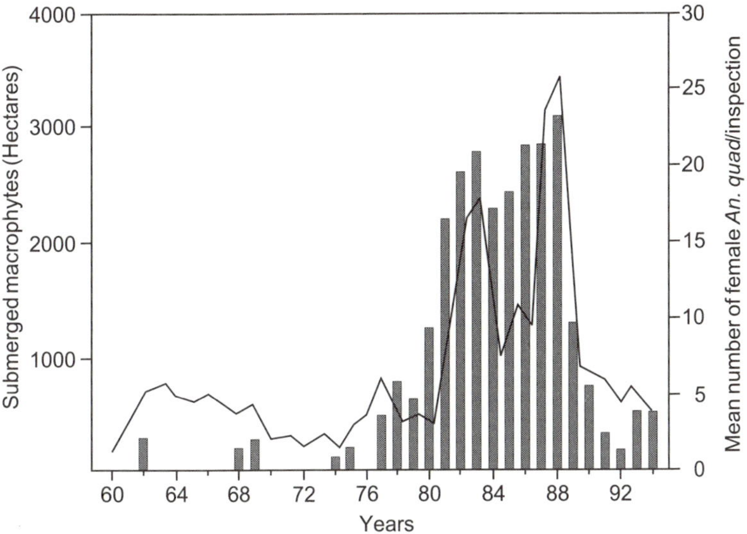

Figure 12.6 Chickamauga mosquito populations and weeds. Bars, submersed macrophytes, line, female *Anopheles quadrimaculatus*.

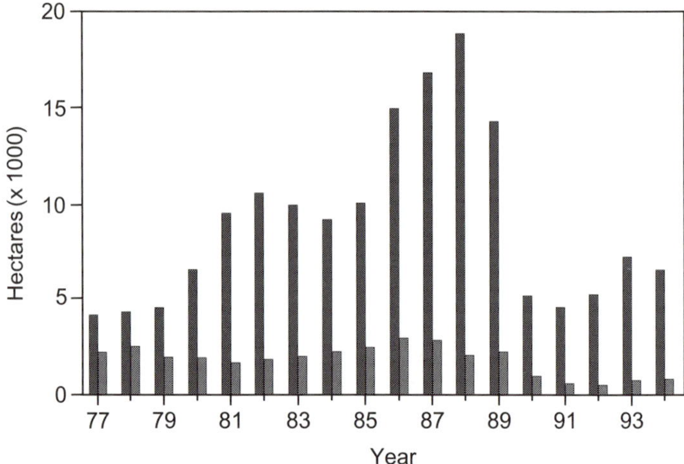

Figure 12.7 Aquatic plants and herbicide usage from 1977 to 1994. Dark bar, aquatic plant hectares; light bar, herbicide treatment.

Milfoil is the predominant macrophyte in the TVA system, but other exotics such as spineyleaf naiad (*Najas minor*) and hydrilla (*Hydrilla verticillata*), which was introduced in the early 1980s, and several native species contribute to total plant coverage. The level of colonization has fluctuated in response to environmental conditions. For example, with an extended drought from 1984 to 1988, the area of aquatic plants doubled from 9,300 to 18,600 ha (Figure 12.7). Clear water and low flows associated with the drought were ideal for submersed aquatic plant growth. By contrast, in 1989, flooding caused a drastic decline. High flows and turbid waters continued in 1990 and 1991, resulting in a decline in plant levels to fewer than 5,665 ha. Since 1991, plant coverage on most reservoirs has increased due to improved growing conditions.

TVA's aquatic plant management programme has relied on the use of Environmental Protection Agency (EPA) approved herbicides to control submersed aquatic plants in designated areas along developed shorelines. This has conferred extra benefits in terms of mosquito control. However, since herbicide use has been deliberately minimized for application to designated areas where it conflicts with multipurpose reservoir uses, extensive plant colonies have persisted each year to provide abundant mosquito habitat. This has meant that the area treated with larvicide for mosquito control may reach 16,188 ha in any year.

The introduction of *Hydrilla* into the Guntersville Reservoir in northern Alabama in the 1980s caused increased concern because it is more competitive than other exotic and native species. It requires less light and will grow in deeper water than milfoil. While it may not be as suitable for mosquito habitat, it has the potential to become the dominant aquatic macrophyte in the TVA reservoir system and could colonize up to 11,736 ha on Guntersville Reservoir alone.

In 1989, in a joint effort with the US Army Corps of Engineers, TVA initiated research and demonstration projects evaluating biological controls for aquatic plants. This effort was initiated to help diminish the use of herbicides and insecticides in TVA reservoirs. A major effort has been directed against *Hydrilla verticillata*. One promising agent is the Indian leaf-mining fly *Hydrellia pakistanae*, which was imported to the TVA rearing facility at Muscle Shoals in 1991. It was first released during the summer of 1991, and since that time nearly 600,000 individuals have been released at several Tennessee Valley locations. *Hydrellia pakistanae* was released in the Guntersville Reservoir during the summer of 1995. This fly is now permanently established throughout southern Florida and the south-eastern USA. At least 70 per cent of the release attempts have been successful as judged by assessment from four months post-release (Grodowitz and Snoddy 1992).

Other biological agents tested to date include the Australian stem-borer weevil *Bagous hydrillae* and the Australian hydrilla leaf-miner *Hydrellia balciunasi*. These biocontrol agents suppress aquatic plants that serve as mosquito habitat. These agents may become valuable alternatives for mosquito control programmes.

Another noxious plant that has been discovered in the Tennessee Valley is purple loosestrife (*Lythrum salicaria*). The native leaf beetle *Altica* sp. and old world exotic leaf feeders (*Galerucella pusilla and G. calmariensis*) are currently being evaluated as control agents for loosestrife and appear promising.

The grass carp (*Ctenopharyngodon idella*), also commonly known as the white amur, is a herbivorous fish native to large rivers of eastern Asia. Although cultured as food fish in Asia and eastern Europe (Miller and Decell 1984), its introduction into the USA in the 1960s was primarily for aquatic weed control (Sutton 1977). It has been used extensively in the south-eastern USA for aquatic plant control in small closed systems (ponds, reservoirs, natural lakes) and for control in irrigation canals in southern California (Stocker and Hagstrom 1986).

Grass carp selectively feed on preferred plants before consuming less palatable species (Leslie *et al.* 1987; Miller and Decell 1984). From TVA studies (TVA 1987; 1989; Webb *et al.* 1994) grass carp stocked at rates of 20–47 fish/vegetated ha almost totally eliminate and suppress submersed aquatic plants such as *Hydrilla*, naiads (*Najas minor* and *Najas guadalupensis*), pondweeds (*Potamogeton* spp.), and muskgrass (*Chara* spp.). Although reductions in cover and biomass of Eurasian water milfoil were observed where grass carp were stocked at 47 fish/vegetated ha, this vegetation was not suppressed below nuisance level in all sections of the test area. However, mosquito population densities decreased significantly. At stocking rates of 75–100 grass carp/vegetated ha, Eurasian water milfoil can also be suppressed (Leslie *et al.* 1987; Martyn *et al.* 1986; TVA 1987; 1989; Van Dyke *et al.*, 1984). However, Leslie *et al.* report that it has been difficult to achieve intermediate levels of aquatic weed control using grass carp in the USA since their usage may result in total removal of all aquatic vegetation. The level of aquatic vegetation control that can be achieved with grass carp is dependent on a variety of factors: the stocking rate, the species of aquatic plants, the climatic regime, the size of a waterbody, and whether the body is an open or closed system.

Grass carp are known to migrate long distances. If stocked in open systems, the fish may move to areas where its herbivorous habits might cause a problem (e.g. rice fields). Even when they are stocked in small ponds and lakes, barriers are generally recommended to prevent their escape during flood events. Natural reproduction may also occur in large free-flowing rivers, thus upsetting the ecological balance of the entire system. Techniques for production of sterile (triploid) grass carp can eliminate this possibility (Cassani and Canton 1985). Smith and Shireman (1983) provide a summary of the biology and use of grass carp for aquatic plant control. Guidelines for various types of waterbodies can be found in Sutton and Vandiver (1986), Leslie *et al.* (1987), Wiley *et al.* (1987), and Bain (1993).

12.2.7 Water level management

Water level management can be used as a very effective tool for controlling permanent pool types of mosquitoes on impounded waters. The management scheme can be a simple continuous lowering of the pool level throughout the breeding season so that water is continually being drawn away from encroaching marginal vegetation (Figure 12.8). This type of operation works well on reservoirs that have a dual purpose (flood control and storage for downstream uses) since normal operation of these type of reservoirs (on tributaries) involves

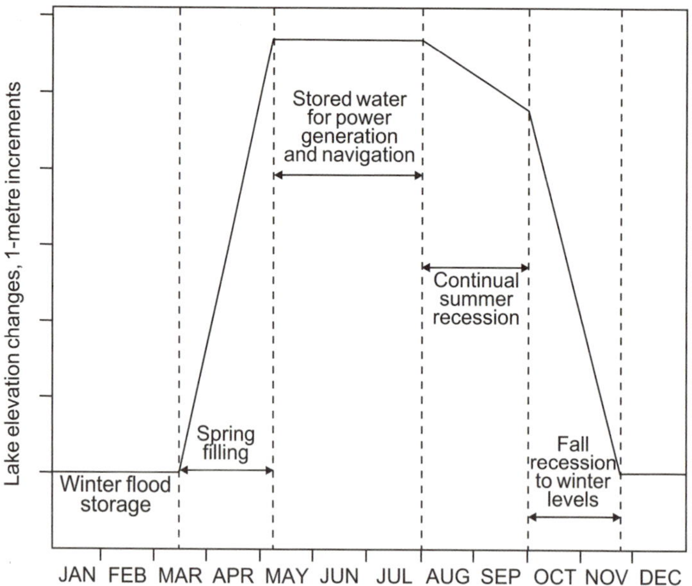

Figure 12.8 Continuous draw of reservoir throughout a typical year in the Tennessee Valley.

a steadily receding water level from mid-summer through to autumn. However, a recent evaluation of water level management by the TVA (1990), published as part of its Lake Improvement Plan, was prompted in part by customers requesting different lake levels for other purposes. The basic changes involved holding summer pool levels longer and delaying the start of the seasonal recession. Consequently, relatively static pool levels ensure continuous contact with aquatic vegetation, thereby creating extensive mosquito larval habitat. Additional remedial control operations have been required as a result of these changes.

Water level management remains the single most important mosquito control measure used by TVA to control permanent pool mosquitoes. However, its effectiveness has been diminished in recent years with the introduction of submersed exotic macrophytes i.e. Eurasian water milfoil and *Hydrilla* into the reservoir system. For many years prior to the introduction of these exotic species, fluctuating water levels provided excellent control of mosquitoes on many reservoirs where the breeding habitat was limited to a relatively narrow band of marginal emergent vegetation. During the peak breeding season on mainstream reservoirs (June to September), eggs, larvae, and pupae were controlled by abruptly lowering the water level by 0.3 m followed by a return to summer pool levels within the same week (Figure 12.9). This abrupt dewatering strands the immature mosquitoes on dry land or exposes them in open water to natural predators. However, breeding habitat created by the submersed aquatic plant

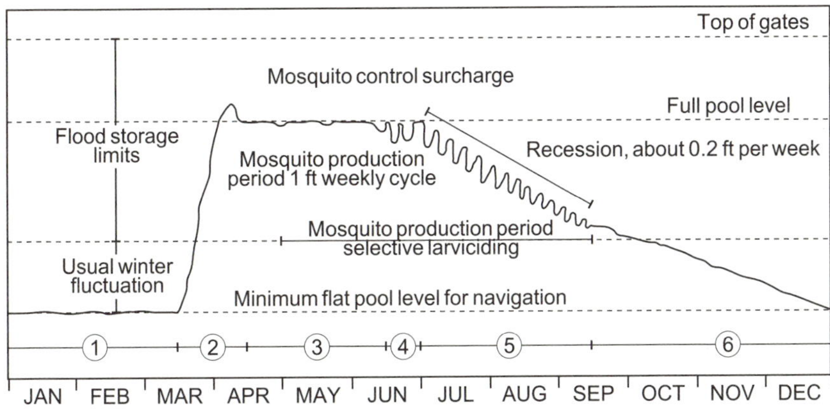

Figure 12.9 Cyclical fluctuation: 1 Low winter flood control levels, controls growth of submerged aquatics, permits marginal drainage and herbicidal operations. 2 Early spring filling, retards plant growth, surcharge-strands drift above full pool, occasionally planned. 3 Constant level pool, provides long-range plant growth control. 4 Cyclic fluctuation, destroys mosquito eggs and larvae, reduces breeding areas, provides clean shoreline. 5 Fluctuation and recession destroy eggs, larvae and pupae, reduce breeding areas and provide clean shoreline. 6 Recession to winter levels, permits fall shoreline maintenance and improvement operations.

species is not similarly impacted, except for those plants growing in the fluctuation zone. Eurasian water milfoil and *Hydrilla* can grow in water that is more than 3 m deep, although the normal winter lake level recession generally limits growth to approximately 2 m.

Extended summer drawdowns at Guntersville Reservoir were designed to examine the control possibilities on Eurasian water milfoil. However, it was discovered that this action had a lasting impact on mosquito production. The proposal to draw the lake level 0.76 m down from its normal summer pool elevation was designed to take advantage of the hot desiccating conditions of mid-summer to affect exposed plants at peak seasonal growth before seed production occurred. It would have been desirable from an aquatic plant management perspective to go even another 0.76 m lower but other programme concerns (water supply, navigation, etc.) precluded this possibility. The lake level was to remain at the minimum elevation of 180.35 m above mean sea level (MSL) for about one week and then be restored to the normal pool elevation of 181.05 m (MSL). Normal pool level for Guntersville Reservoir at this time of year drops from 181.35 m to 181.05 m (MSL).

The year 1982 was typical for anophelines (Figure 12.10). In 1983, a similar trend began but was interrupted by the summer drawdown. The impact on larval population densities was of such a magnitude and duration that larvicidal applications were all but eliminated from this time to the end of the season. The drawdown essentially eliminated most submersed aquatic vegetation located around the perimeter to a depth of about 0.6 m in the reservoir, and in doing so eliminated prime larval habitat.

Water level management schemes can be developed and implemented in a variety of ways to manage aquatic plants and mosquitoes, directly or indirectly, as long as mechanisms exist for fluctuating water levels and operating schemes are compatible with other programme interests.

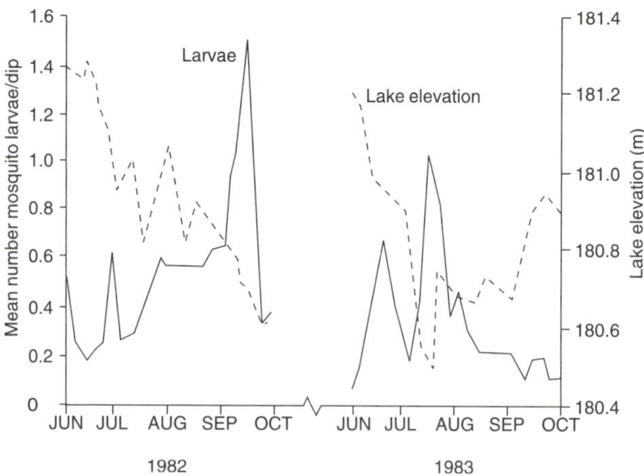

Figure 12.10 Summer drawdown, Tennessee Valley reservoirs.

12.2.8 Use of mosquito larvicide

Larvicide is only applied in critical areas to supplement physical and naturalistic measures for controlling mosquitoes. The need for larviciding varies from season to season and from week to week. Larviciding is also determined by shoreline conditions, expected water levels, prevailing weather, human activity, and current mosquito production. Current mosquito production is measured by results from some 150 strategically located red wooden boxes which are open on one side to provide resting places for adult mosquitoes. Other types of adult mosquito resting areas may also occasionally be inspected to obtain a quick index of mosquito populations in unmonitored areas.

If adult mosquitoes are present at an index station, a comprehensive larval survey is conducted of nearby breeding habitat in the associated reservoir marginal area using the standard dipper technique. If the larval index exceeds an average of 0.25 larvae per dip, larvicidal treatments are applied. The only mosquito larvicide presently used operationally for controlling permanent pool species of mosquitoes is *Bacillus thuringiensis israelensis* (*Bti*) (Table 12.1). This commercially produced bacterial agent is applied by helicopter to breeding habitat at a rate of 585 ml of active ingredient/ha. Applications may be repeated as many as eight times in specific areas, primarily on seven reservoirs. The total area treated annually may exceed 16,000 ha. TVA's programme uses only EPA-approved compounds, in accordance with label specifications, that have been thoroughly reviewed and tested. We monitor their operational usage and run a supplementary programme to evaluate other promising compounds for their potential use.

12.2.9 Use of mosquito adulticide

When the desired level of control has not been achieved through the use of the other methods described, mosquito adulticidal compounds are applied. They are applied in a cold aerosol formulation at an ultra low volume (ULV) rate of 118 ml of finished material per minute or 47.8 ml/ha. The aerosol is applied from truck-mounted generators (Leco® models) at times when adult mosquitoes are most active and climatic conditions are favourable. For maximum control of most species, applications are made from dusk until about three hours afterwards.

Three EPA-registered compounds are presently used in this work: resmethrin in the manufactured formulation Scourge®; malathion identified in the formulation Fyfanon®, and permethrin in the trade name Permanone®. Applications are made primarily in residential and public use areas within flight range (1.6 km) of permanent pool mosquitoes that have bred on TVA lands and waters. Scourge is the primary compound used for control of permanent pool species while malathion and permethrin are used sparingly. Scourge has replaced malathion in recent years due to the development of widespread resistance in permanent pool species. However, malathion is still used occasionally for the control of susceptible populations of floodwater mosquitoes.

Table 12.1 Labelled insecticides and dosages in controlling adult and larval mosquitoes in the Tennessee Valley region

Mosquito breeding habitat	*Target species*	*Vector status, actual and potential*	*Pest status*[a]	*Insecticides used*	*Application rate*	*Application method*
				Adulticides		
Reservoir impoundment	*Anopheles quadrimaculatus*	Malaria, canine filariasis	U	1 Malathion 91% (Fyfanon®)	88–118.3 ml/min[b]	Ultra low volume by cold fog generation
(Permanent pool)	*An. punctipennis*	Canine filariasis	U	2 Resmethrin 4% (Scourge®)[d]	117.5–136.6 ml/min[b,c]	Ultra low volume by cold fog generation
			U	3 Permethrin 4% (Permanone®)[e]	147.9 ml/min[b,c]	Ultra low volume by cold fog generation
				Larvicides		
	Culex erraticus		U	1 *Bti* H14 1200 ITU (Vectobac 12AS®)	0.6–1.2 l/ha	Aerial, ground, or boat application
				2 Temephos 43% (Abate 4E)	9.4–27.6 ml/ha	Aerial, ground, or boat application
Reservoir flood zone				*Adulticides*		
	Aedes atlanticus	Californian encephalitis	I	1 Malathion 91% (Fyfanon®)	88.7–118.3 ml/min[b]	Ultra low volume by cold fog generation
	Ae. sticticus	Canine filariasis	C			
(Floodwater and transient pool)	*Ae. trivittatus*	Californian encephalitis	I	2 Resmethrin 4% (Scourge®)[d]	177.5–236.6 ml/min[b,c]	Ultra low volume by cold fog generation
	Ae. vexans	Canine filariasis and eastern equine encephalitis	C			
				3 Permethrin 4% (Permanone®)	147.9 ml/min[b,c]	Ultra low volume by cold generation

	Psorophora ciliata	Venezuelan equine encephalitis	U	*Larvicides* 1 *Bti* H14 1200 ITU (Vectobac-G®)	2.8–11.2 kg/ha	Aerial, ground, boat application, cyclone seeder, power blower, etc.
	Ps. columbiae		C			
	Ps. cyanescens	St Louis encephalitis, canine filariasis	C	2 Temephos 2% (Abate® 2G)	5.6 kg/ha	Aerial, ground, boat application, cyclone seeder, power blower, etc.
	Ps. varipes *Ps. varipes*					
Artificial containers	*Cx. pipiens-quinquefasciatus*	Canine filariasis, St Louis encephalitis	C	*Adulticides* 1 Malathion 91% (Fyfanon®)	88.7–118.3 ml/min[b]	Ultra low volume by cold fog generation
	Cx. restuans	Canine filariasis, western equine encephalitis	U	2 Resmethrin 4% (Scourge®)[d]	177.5–236.6 ml/min[b,c]	Ultra low volume by cold
	Ae. aegypti *Ae. triseriatus*	Yellow fever, canine filariasis, and dengue	I	*Larvicides* 1 Chlorpyrifos 22% (Dursban® 2E) 4%	116.9–233.9 ml/ha	Hand sprayer and power sprayers
		California encephalitis, canine filariasis	C	2 Temephos 43% (Abate® 4E)	9.4–27.6 ml/ha	Hand sprayer and power sprayers

[a] Pest status: U, usually; I, infrequently; C, constantly.
[b] Vehicle speed of 16.1 km/h.
[c] Mixture in light mineral oil (54 second viscosity) or comparable solvent. See manufacturer's recommendations.
[d] Scourge® – Penick Corporation trade name for mixture of 18% W/W resmethrin + 54% piperonyl butoxide.
[e] Permanone® – AgrEvo Corporation trade name for mixture of 31% W/W permethrin + 66% piperonyl butoxide.

Acknowledgments

Sincere thanks are extended to William K. Oldland II and Jason P. Allen for their help in the preparation of this chapter and useful data utilized herein, Kenneth J. Tennessen for his help in preparing the figures for this chapter, Earl R. Burns for supplying updated herbicide data, David H. Webb for supplying information pertaining to the grass carp, and A. Leon Bates for supplying plant population data.

References

Bain, M.B. (1993) Assessing impacts of introduced aquatic species: grass carp in large systems. *Environmental Management*, **17**, 211–224.

Cassani, J.R. and Canton, W.E. (1985) Induced triploidy in grass carp, *Ctenopharyngodon idella* Val. *Aquaculture*, **46**, 37–44.

Gartrell, F.E., Cooney, J.C., Chambers, G.P. and Brooks, R.H. (1981) TVA Mosquito Control 1934–1980 – Experience and Current Program Trends and Developments. *Mosquito News*, **41**, 302–322.

Grodowitz, M.J. and Snoddy, E.L. (1992) Release and establishment of insect biocontrol agents of *Hydrilla* in Alabama, Louisiana, and Texas. *Proceedings, 25th Annual Meeting, Aquatic Plant Control Research Program.* Miscellaneous Paper A-92-2, U.S. Army Engineer Waterways Experiment Station, Vicksburg, Mississippi. pp. 226–233.

Hammer, D.E. and Kadlec, R.H. (1983) Design principles for wetland treatment systems. *U.S. EPA Report*, PB-83-188-722.

Leslie, A.J., Van Dyke, J.M., Hestand, R.S. and Thompson, B.Z. (1987) Management of aquatic plants in multi-use lakes with grass carp *(Ctenopharyngodon idella). Lake and Reservoir Management* **3**, 266–276.

Martyn, R.D., Noble, R.L., Bettoli, P.W. and Maggio, R.C. (1986) Mapping aquatic weeds with aerial color infrared photography and evaluating their control by grass carp. *Journal of Aquatic Plant Management*, **24**, 46–56.

Miller, A.C. and Decell, J.L. (1984) *Use of the white amur for aquatic plant management.* Instruction Report A-84–1, U.S. Army Engineer Waterways Experiment Station, Vicksburg, Mississippi.

Smith, C.R. and Shireman, J.V. (1983) *White amur bibliography.* Miscellaneous Paper A-83-7, prepared by the University of Florida, Gainesville for the U.S. Army Engineer Waterways Experiment Station, Vicksburg, Mississippi.

Snoddy, E.L. and Cooney, J.C. (1989) Pesticides in terrestrial and aquatic environments. *Proceedings of a National Research Conference, May 11–12, 1989.* Virginia Water Resources Research Center, Virginia Polytechnic Institute and State University, Blackburg, Virginia. pp. 440–443.

Stocker, R.K. and Hagstrom, N.T. (1986) Control of submersed aquatic plants with triploid grass carp in southern California irrigation canals. *Lake and Reservoir Management*, **2**, 41–45.

Sutton, D.L. (1977) Grass carp (*Ctenopharyngodon idella* Val.) in North America. *Aquatic Botany* **3**, 157–164.

Sutton, D.L. and Vandiver Jr, J.J. (1986) Grass carp – A fish for biological management of *Hydrilla* and other aquatic weeds in Florida. Bulletin 867. Florida Agricultural

Experiment Station, Institute of Food and Agricultural Services, University of Florida, Gainesville.

Tennessee Valley Authority (1987) *White amur project*. Tennessee Valley Authority, Office of Natural Resources and Economic Development, Division of Air and Water Resources, TVA/ONRED/AWR-87/38. pp. 383.

Tennessee Valley Authority (1989) *Supplement to white amur project report*. Tennessee Valley Authority, Resource Development, River Basin Operations, Water Resources, TVA/WR/AB-89/1. pp. 82.

Tennessee Valley Authority (1990) *Tennessee River and Reservoir System Operation and Planning Review. Final Environmental Impact Statement*. TVA, Knoxville.

Tennessee Valley Authority and the United States Public Health Service (1947) *Malaria Control on Impounded Water*. US Government Printing Office, Washington, DC.

Van Dyke, J.M., Leslie Jr, A.J. and Nall, L.E. (1984) The effects of the grass carp on the aquatic macrophytes of four Florida lakes. *Journal of Aquatic Plant Management*, **22**, 87–95.

Webb, D.H., Mangum, L.N., Bates, A.L. and Murphy, H.D. (1994) Aquatic vegetation in Guntersville Reservoir following grass carp stocking. *Proceedings of Grass Carp Symposium*, University of Florida, Gainesville, Florida. pp. 199–209.

Wiley, M.J., Tazik, P.P. and Sobaski, S.T. (1987) Controlling aquatic vegetation with triploid grass carp. *Illinois Natural History Survey, Champaign*. Circular 57. pp. 1–16.

Chapter 13

Aquaculture, anophelines and environmental solutions from Indonesia

W. Bart Snellen

13.1 Introduction

As one of its activities as a Collaborating Centre of the WHO/FAO/UNEP/UNCHS Panel of Experts on Environmental Management for Vector Control (PEEM), the International Institute for Land Reclamation and Improvement (ILRI) reviewed literature on malaria control in Indonesia prior to 1940 when it was a colonial administration of the Dutch East Indies. This review had special emphasis on measures related to land and water development.

This chapter describes how aquaculture was first promoted because of the extra food, income and health that it would provide. The policy was later condemned as a fatal error, on the grounds that it caused widespread malaria. It also describes how a solution was arrived at: an environmental management method that proved successful in controlling malaria without having to abolish fish cultivation. This of course has ramifications for other countries, e.g. Thailand, where aquaculture has become an important revenue earner as well as a source of extra protein.

13.2 Marine fishponds: source of income and malaria

The cultivation of fish in artificial ponds in the coastal zone has been practised for centuries in northern Java. The earliest reference is a Javanese code of law from around AD 1400, which specifies punishment for stealing from a marine fishpond.

In 1864 the agricultural inspector Van Spall investigated both the fresh and salt water fishpond cultivation practices of local farmers. Marine fishponds in Java at that time occupied an area of 33,000 ha. Van Spall recommended that the government should actively support the expansion of the cultivated area by transforming mangrove forest into fishponds. This would provide extra food, income and tax revenue, and improve health.

By 1926, marine fishponds occupied an area of 55,000 ha, producing more than 17 million kilograms of fish with a market value of almost DG7 million. The Health Service was not at all happy with the expansion, however. In a manual for government officials on malaria matters, Rodenwaldt (1925) wrote:

> Concerning the large population centres on Java's northern coast, we are confronted with a fatal error. Here one has propagated the establishment of fishponds in the mangrove coastal zone, which was held responsible for malaria, in the conviction that by doing so the zone would be made safe. At present we know that *Anopheles sundaicus*, the most dangerous malaria vector of the Netherlands–Indies, breeds in these fishponds which are therefore the cause of heavy endemic malaria in coastal settlements. It is the task of the hygienist to try and eliminate these marine fishponds from the surroundings of human settlements. The matter is further complicated by the fact that *Anopheles sundaicus* has been found to be capable of flying very long distances, which means that all fishponds within at least 3 km from human settlements need to disappear. Tens of years will be needed to accomplish this task and even then, it remains to be seen whether in certain regions economic considerations will not prevail over hygienic ones. One lesson however has to be learned from the results of biological and hygienic studies of the last ten years: that in the Netherlands–Indies under no circumstances can it be permitted that new fishponds are established in the coastal zone, or ponds that have been given up are taken into production again.
>
> (Rodenwaldt 1925)

In spite of Rodenwaldt's strong objections, the area of fishponds increased further from 55,000 ha in 1926 to 82,000 ha in 1941. We will examine how opinions on the relationship between marine fishponds, malaria and its control changed over the years.

13.3 Extensive breeding of anopheline mosquitoes

In a malaria survey in the capital, Jakarta, in the second half of 1907, the medical researcher Kiewiet de Jonge (1908) found that the mortality rate among the local population in the district of Pendjaringan near the coast was three to four times higher than that in the other districts of the city. Also, the percentage of people with enlarged spleens was much higher in the coastal districts.

Kiewiet de Jonge then searched for breeding sites of anopheline mosquitoes. He found them almost everywhere: in rice fields in all parts of the capital (which was in fact more a collection of villages), in the marine fishponds, in numerous puddles, wheelruts, hoofprints, near wells and in wash-places, so that he exclaimed: 'Yes, where didn't we find anopheline breeding sites!'

In his report, Kiewiet de Jonge (1908) considered the following control measures:

- elimination of all breeding sites;
- construction of mosquito proof houses;
- evacuation of the unhealthy town districts;
- drug treatment of malaria patients.

Kiewiet de Jonge considered the first three options to be unfeasible. He recommended that the local administration set up a programme for the distribution of quinine, free of charge, to all malaria patients. The administration provided funds and made Kiewiet de Jonge responsible for the implementation of the programme. From his report of 1908, it is apparent that Kiewiet de Jonge considered the programme to be a temporary solution at best, and also that he was rather disappointed to find that the population responded with indifference to his efforts.

13.4 Jakarta sanitation project 1913

In 1913, the Public Works Department began an ambitious sanitation project in the capital. Since its foundation in 1610, Jakarta had been an unhealthy place. The Dutch founders had considered the location amidst swamps a strategic advantage. After Dutch fashion, a network of canals was constructed in the mid-seventeenth century.

In 1699, there was a large volcanic explosion of Mount Salak, 60 km south of Jakarta. Subsequent floods carried enormous amounts of eruption material, which blocked the canals and drains and created evil-smelling swamps. In the eighteenth century, several attempts to improve the situation by diverting floodwaters away from the city failed. Early in the nineteenth century, a new adminstrative and residential area was built on higher ground, in order to escape the poor hygienic conditions in the old city. After the exodus of the Europeans, the sanitary conditions in the lower city deteriorated even further. By 1910, three centuries after its foundation, the health conditions were considered unacceptable.

The estimated cost of the sanitation works that began in 1913 was DG4 million. The major components were:

- flood diversion works;
- construction of a drainage and sewerage system;
- ground water level control;
- elimination of stagnant water.

Although the works would reduce the number of mosquito breeding sites, the plan was not specifically aimed at malaria control. The rice fields and the marine fishponds, in which Kiewiet de Jonge had found anopheline larvae, would remain unaffected.

13.5 Marine fishponds: breeding sites of malaria vectors

Five years later, the chief designer of the 1913 sanitation plan made an urgent request to the administration for allocation of an additional DG1.6 million for the expropriation and reclamation of the fishponds in the coastal zone of Jakarta, and to allocate a sum of DG25,000 annually over a period of 30 years, for pumping costs and gradual filling of the area with sludge from dredging operations in the Jakarta harbour and waterways.

This change of ideas was brought about by the outcome of the malaria investigations by van Breemen and Sunier in 1917 and 1918. The following is based on a translated summary of their research conclusions, which were an integral part of the advice of the Health Committee to the City Council of Jakarta (Anonymous 1919):

> In the summary below we briefly indicate what we at present know for certain about the relationship that exists here in Batavia [Jakarta] between the marine fishponds and the spread of malaria.
>
> 1 In Batavia spleen index and mortality are highest in the marine fishpond zone and gradually decrease from there to the south [Figure 13.1].
> 2 Larvae of the very dangerous malaria transmitting mosquito *Myzomyia ludlowi* Theobald [*Anopheles sundaicus*] were found, with very few exceptions, in the marine fishpond zone only and mainly in the fishponds themselves.
> 3 Breeding sites of other anophelines, which do not play a role here, were found all over Batavia. In these breeding sites, however, no *M. ludlowi* larvae were found.
> 4 In locations south of the fishpond zone with high spleen indices, where many *M. ludlowi* mosquitoes were caught in the houses, we could not find any *M. ludlowi* breeding sites.
> 5 The fish cultivated in the marine fishponds (*Chanos chanos*) [Figure 13.2] is vegetarian and feeds on submerged waterplants which grow close to the surface. These waterplants grow best when salinity is around 20 parts per thousand; growth decreases at higher salinity levels.
> 6 In addition to the cultivated fish, the marine ponds often contain enormous quantities of a small fish (*Haplochilus panchax*) [Figure 13.2] which in the literature on malaria control is mentioned as one of the best destroyers of mosquito larvae.
> 7 In spite of these larvae-eating fish, the ponds which contain the water-plants mentioned under 5 – and also the overgrown banks of the ponds – produce a large number of mosquitoes, varying from a few up to several hundred per square metre. In the absence of larvae-eating fish, mosquito production figures as high as 6000/m^2 per night have been observed.
> 8 The production of mosquitoes decreases with increasing salinity and stops completely when salinity in the ponds is considerably higher than seawater.
> 9 The level of salinity at which mosquito production stops is so high that it is not even reached in a year with a less pronounced dry season. This indicates that stopping mosquito production through maintaining a high salinity level in the ponds is not feasible.

> ... From the above it should be clear for everyone that the actual source of heavy endemic malaria in Batavia is located exclusively in the brackish water zone and specifically in the marine fishponds.
>
> (Anonymous 1919: 223–225)

The researchers, and with them the Health Committee, recommended the following to the City Council of Jakarta:

> Short of moving Batavia to a location outside the sphere of influence of the brackish water zone, there is only one permanent and effective solution: to fill in the fishponds and reclaim and bring under cultivation the entire brackish water zone north of Batavia.
>
> (Anonymous 1919: 223–225)

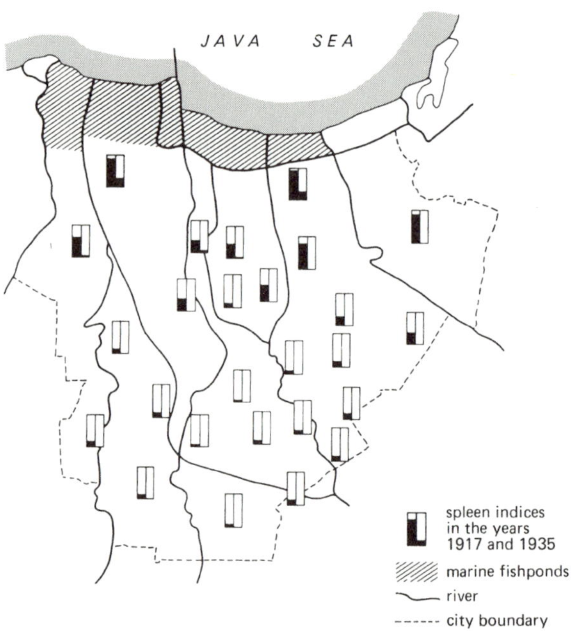

Figure 13.1 Map of Jakarta, indicating location of marine fishponds and the spleen indices for the various quarters in the years 1917 and 1935.

In their summary report, the researchers indicated that various parties had asked whether it would not be possible to exploit the fishponds in such a way that they were unsuitable as breeding sites for the malaria vector, for instance, by removal of water plants near the surface, which provided shelter from larvae-eating fish. The difficulty involved here, as pointed out by the researchers, was that the cultivated fish depend on these water plants for adequate growth. They mentioned

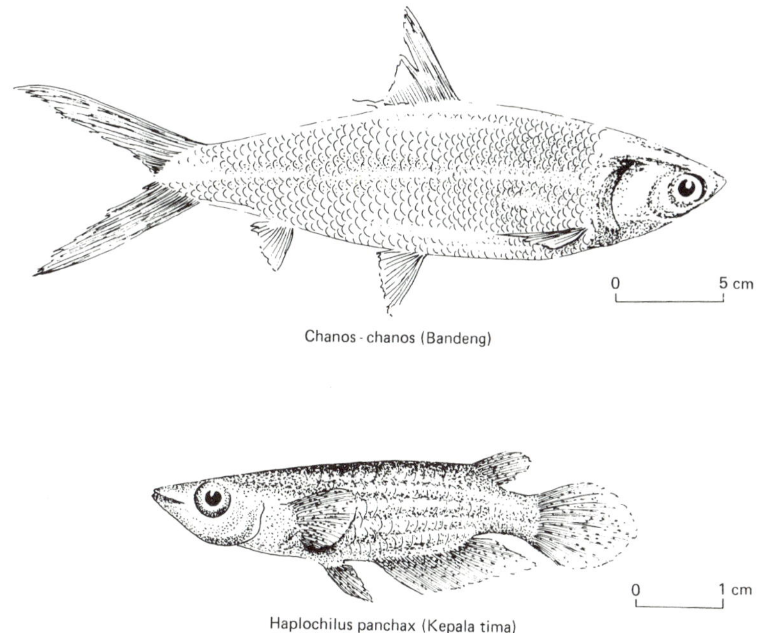

Figure 13.2 *Chanos chanos*, the cultivated fish in the marine fishponds, and *Haplochilus panchax*, an effective larvae-eating fish. (Redrawn by Betty van Aarst, International Land Reclamation and Improvement Institute; based on Figure 21 in Kuipers 1943.)

the idea of conducting trials on artificial feeding of the cultivated fish, yet warned strongly against such attempts:

> Even if such a way of exploitation were technically possible, it remains questionable whether fishpond owners could be made to adopt the new methods. And apart from that, exploitation of the marine fishponds in such a way that they no longer present a danger to public health would require permanent close supervision. . . . the undersigned are of the opinion that from a public health point of view such attempts to 'run with the hare and hold with the hounds' are totally unacceptable. Only a radical and permanent clearing of all the water collections in the brackish water zone can save the population of Batavia from the paralysing pressure of endemic malaria with certainty and forever.
>
> (van Breemen and Sunier, cited in Anonymous 1919: 223–225)

The words of the researchers did not fail to make an impression on the City Council of Jakarta. They adopted the proposal for a radical and permanent solution and requested the government provide funds to implement the works as part of the sanitation programme that had started in 1913.

The government was slow to react and perhaps not without reason: not only was implementation of the plans very expensive, it also meant loss of a popular source of high protein foods, loss of income and job opportunities for the owners of the ponds, and loss in tax revenue. It was not until 1928 that a sum of any significance became available for malaria control in the marine fishponds in Jakarta. The time lag provided the opportunity for the initiators and those opposed to their plans to get together.

13.6 An appeal from east Java

Reijntjes, inspector of the inland fisheries department in Surabaya, east Java, wrote an article on malaria and fishponds in the agricultural journal *Teysmannia* in 1922. Reijntjes disagreed entirely with the researchers in Jakarta on the crucial issue of the type of plants that cultivated fish feed on (Figure 13.3). As indicated above, the researchers in Jakarta claimed that the waterplants floating near the surface, which provided shelter from the larvae-eating fish, were essential for

Figure 13.3 Collecting mosquito larvae. (Redrawn by Betty van Aarst, International Institute for Land Reclamation and Improvement, from photographs in Schüffner and Swellengrebel 1918.)

adequate growth of the cultivated fish. Hence, their conclusion that fish cultivation and breeding of malaria mosquitoes go hand in hand. Reijntjes, however, conducted experiments that indicated that the cultivated fish fed mainly on the blue–green algae that developed at the bottom of the ponds. He grew waterplants in one of two otherwise similar ponds and then stocked the ponds with an equal number of fish. After two months, the fish in the pond without floating plants were larger, while the number of fish in the pond with the waterplants doubled. The mass of waterplants remained untouched. After two months there was still an abundance of waterplants but the fish had lost weight.

Reijntjes pointed out that in east Java, fish were cultivated in ponds with hardly any plants at the water surface. The common practice there was to drain the ponds for a few days every month to expose the bottom to sunlight, which stimulated the development of the bottom algae and killed the floating waterplants. Subsequently, the ponds were filled with sea water. By applying this method in Jakarta, Reijntjes thought that it would be possible to combine the interests of fish cultivation and malaria control. The method depends on adequate water management in the ponds, which require a system of drainage and supply canals. As a first step, Reijntjes recommended trials on improved water management in the fishponds of Jakarta.

Unfortunately, the recommendations from east Java took a long time to reach the researchers in Jakarta and it took six years before these trials were begun.

13.7 Experiments in pond water management

Unaware of the practices in east Java and lacking funds for the implementation of the radical and permananent solution they favoured, the malariologists in Jakarta started looking for less costly control measures; later on they initiated the very type of trials that they had warned against earlier.

A low cost method was to give up fish production and simply cut away a few metres of the dykes of the fishponds, providing free access to the sea water. The tidal movement then would render the ponds unsuitable as breeding sites for the malaria vector. Mangkoewinoto had successfully applied this method in neglected fishponds at Probolinggo (east Java) in 1921 (described together with van Breemen's experiments in pond water management in van Breemen 1930).

Van Breemen tried this method on a fishpond complex at Tandung Periuk, the harbour of Jakarta, in 1923 and again in 1924, with very poor results. It appeared that the tidal amplitude at Jakarta, which is about 1 m at spring tide compared with 2.5–3 m in east Java, was too small to suppress the production of floating algae and therefore the breeding of mosquitoes.

After this failure, van Breemen conducted an experiment on 10 ha of fishponds in which he achieved a clean water surface by overstocking the ponds. While the productivity of the ponds was lower than usual, the experiment demonstrated that it was possible to produce fish without producing mosquitoes.

The leader of the Central Malaria Bureau at that time was Rodenwaldt, who was very much opposed (as van Breemen had been) to any form of 'hygienic exploitation' of the fishponds on the grounds that it would be impossible to supervise such activities adequately.

13.8 Inventory of exploitation methods

During 1927, Walch was appointed leader of the Central Malaria Bureau. He immediately called for an inventory of the various methods of fishpond exploitation in the archipelago, and their effects on malaria. This study first brought into contact the malariologists from Jakarta and the fisheries inspector Reijntjes from Surabaya.

Reijntjes took them to the district of Pasuruan (east Java), which represented the type of 'hygienic exploitation' he had described in his article of 1922. They witnessed how the mass of floating green algae in the centre of the ponds turned into whitish powder after the ponds had been drained and exposed to sunlight for a day or two, while the fish remained in a 0.3 m deep and 1.5 m wide ring channel. (The researchers from Jakarta were quite impressed: in the course of their own

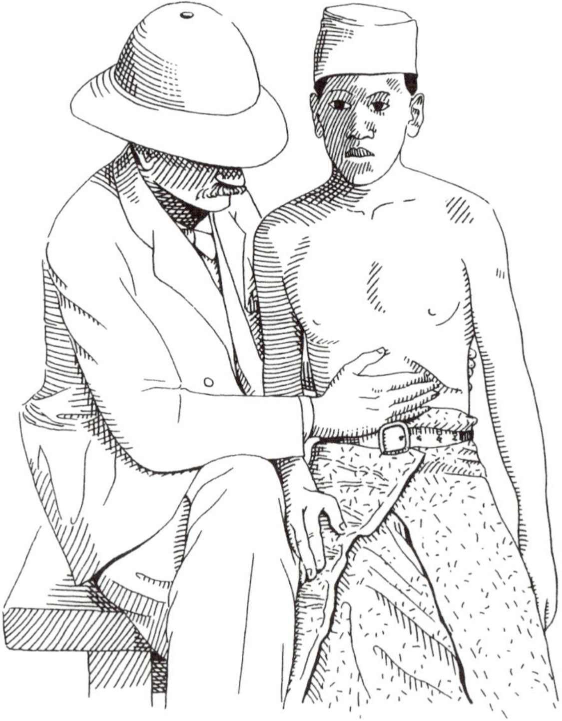

Figure 13.4 Spleen survey. (Redrawn by Betty van Aarst, International Institute for Land Reclamation and Improvement, from photographs in Schüffner and Swellengrebel 1918.)

attempts to obtain a clean water surface they had once had several hundred men working for two days to clear the green algae from a single pond.) After the green algae were killed, a shallow layer of water was maintained for several days, which, again under the influence of sunlight, provided a suitable medium for the development of blue–green algae on the bottom of the ponds. After this, the ponds were filled again with sea water. The malariologists found the fishponds in the district free of floating algae and the villages full of healthy children. This favourable impression was confirmed by a spleen index survey (Figure 13.4). Higher spleen indices encountered in a few villages could be explained by local imperfections, such as:

- ponds with too shallow a water depth, which allowed regeneration of floating green algae; the remedial action required was deepening of these ponds;
- ponds that were supplied by rivers rather than directly from the sea could often not be drained in the rainy season because of high water levels in the river; this resulted in poor development of blue–green algae on the bottom while production of floating green algae remained unchecked and was even stimulated by the lower salinity levels; here, the remedial action involved digging a combined drainage and supply canal directly to the sea.

The excursion to east Java convinced the malariologists that the 'Pasuruan method' of fishpond exploitation was worth trying in Jakarta.

In Semarang, on the north coast of central Java, the researchers came across another exploitation method: fishponds were not drained monthly, but daily, under influence of the tide. This practice had developed without any guidance from the fisheries department, and also resulted in a water surface that was free of floating algae. The malariologists were quite pleased with their 'discovery'. First because the 'Semarang method' proved that the concept of 'hygienic exploitation' was possible with about the same tidal amplitude as that of Jakarta; and second, because such exploitation was accomplished without any supervision.

The inventory also took the malariologists to Probolinggo and Banyuwangi, both in east Java, where, as a malaria control measure, fishponds had been abandoned to the sea. Their main observations were that this method did not really provide a low cost solution because of the compensation payments that had to be made to the owners of the fishponds; and that it did not even provide a permanent solution because additional canals had had to be dug to ensure sufficient tidal action and continuous surveillance was needed to keep farmers from rebuilding the dykes and resuming fishpond cultivation.

13.9 'Hygienic exploitation' after all

Reijntjes' recommendation of 1922, to try the 'Pasuruan method' in Jakarta on an experimental basis, was finally put in to practice in 1928. A fishpond area of 60 ha due north of the city centre and west of the old harbour canal was selected. The objective of the experiment was explained to the owners of the ponds in a meeting that was also attended by the mayor and the regent of Jakarta. If lower fish production than usual resulted from the new exploitation method, the owners would receive compensation. The owners agreed to co-operate and preparations started in 1928. These involved:

- replacing two narrow, winding and overgrown supply ditches by a 7 m wide canal; the capacity of this canal made it possible to fill the ponds in two days;
- constructing a main sluice at the entrance of the supply canal;
- installating a new intake sluice for each of the 23 ponds;
- levelling the bottom of the ponds and where necessary raising the bottom elevation between 0.1 and 0.3 m above low tide level;
- digging a ring channel in each pond.

The works were executed by the Technical Department of the Ministry of Public Health and completed in two-and-a-half months.

13.10 The trial

The first difficulty was to get the owners to stock their ponds with fish. Normally they would not do so until a fair amount of green floating algae had developed. By 20 November 1928, all twenty-three fishponds were finally stocked. To stimulate development of blue–green algae, the water level was allowed to vary daily with the tide, as in the 'Semarang method'. But after this, the researchers wanted to maintain the water level as high as possible, to supress the development of floating green algae, which they saw as the source of malaria, and which the pond owners perceived as essential fish food.

A publication by Walch *et al.* (1930) provides the details of their dealings with individual owners and their actions in individual ponds. It suffices to say here that by implementing the principles of the 'Pasuruan method' they managed to obtain a water surface that was free of both floating algae and mosquito larvae, while producing a sufficient quantity of fish to make exploitation economically viable.

From the production data up to the end of 1929, the researchers calculated that the first, second and third harvest gave the owners an average return on working capital of 23 per cent, 108 per cent and 189 per cent on a yearly base. In their calculations of working capital, they included an annual rent of Dfl 210 per ha, which was not actually paid by the owners in the experiment. At the time, labour cost about Dfl 1 per day and rice was Dfl 0.10 per kg.

13.11 New sanitation plan

Based on the outcome of the experiment, the researchers recommended sanitation of all fishponds in a 4 km long zone north of Jakarta. In contrast to the experiment, the ponds would have to be expropriated and then rented back to the owners. As most ponds were long and narrow, this made them unsuitable for the digging of ring channels, so reshaping was necessary to obtain suitable ponds. The researchers estimated the cost of sanitation of all 1000 ha of fishponds at Dfl 5.6 million.

Between 1928 and 1932, an area of 291 ha of fishponds was converted at a cost of Dfl 2 million. Due to the worldwide recession, no funds were available for the sanitation of the remaining 700 ha.

13.12 Effects of 'hygienic exploitation' on malaria and fish production

Hygienic exploitation was very successful, as can be seen in Figure 13.1 which depicts the spleen index for the various quarters of Jakarta in the years 1917 and 1935. Quarters closest to the sanitized ponds showed the best improvements. Furthermore, after hygienic exploitation of the fishponds started, the overall mortality rate in Jakarta decreased substantially, from a stable 39 per 1000 between 1925 and 1929 to 27 per 1000 in 1931.

Markus (1941) researched into fish production and reported that while productivity in many fishpond areas under 'hygienic exploitation' remained high (500 kg/ha/year), productivity decreased to as low as 50 kg/ha/year in others. Markus found that high productivity was related to ponds with a black homogeneous soft mud with a high content of organic matter. Low productivity was associated with brown to greyish heterogeneous often cloddy mud low in organic matter. Frequent drying of mud layers appeared to reduce the organic matter content.

Another investigator, Vaas (1947), reported that to improve the productivity of the ponds they were no longer drained completely and fertilized with a compost made out of weeds from the pond and vegetation from the banks. Vaas recommended the planting of leguminous crops on the banks that these be incorporated into the compost as a green manure. He did not indicate how successful these measures were in restoring productivity to the fishponds.

From this one can conclude that environmental management was the main strategy for malaria control in Indonesia in the first half of this century before insecticides such as DDT became available. This case study demonstrates that environmental management was successful in controlling malaria caused by anophelines breeding in marine fish ponds, without having to abolish fish cultivation. Takken *et al.* (1991) describe the Indonesian experience with malaria control and provide many other environmental solutions.

References

Anonymous (1919) Zout- en brakwater vischvijvers als bron voor malariagevaar. *De Waterstaats-Ingenieur*, **5**, 223–225.

Kiewiet de Jonge, G.W. (1908) De malariaverhoudingen to Batavia en haar bestrijding. *Geneeskundig Tijdschrift voor Nederlands-Indie*, **68**, 422–457.

Kuipers, J. (1943) *Sanitation Works in the Indian Archipelago.* Typescript with ink and watercolour drawings produced by Kuipers while a prisoner of war of the Japanese.

Markus, B. (1941) Over de slibbodem van zoutwatervijvers. *Landbouw (Buitenzorg, Java)*, **27**, 891–910.

Reijntjes, E.J. (1922) Vischvijvers en malariabestrijding. *Teysmannia*, **33**, 241–253.

Rodenwaldt, E.R.K. (1925) *Handleiding voor bestuursambtenaren in zake malaria-vraagstukken.* Uitgave van den Dienst der Volksgezondheid in Nederlands-Indië.

Schüffner, W. and Swellengrebel, N.H. (1918) *Handleiding voor het epidemiologisch malaria-onderzoek ten behoeve van ambtenaren bij den burgerlijke geneeskundige dienst*, Albrecht & Co., Weltevreden. Also in: *Annual Report 1987, International Institute for Land Reclamation and Improvement*, Wageningen, The Netherlands, p. 14.

Takken, W., Snellen, W.B., Verhave, J.P., Knols, B.G.J. and Atmosoedjono, S. (1991) *Environmental measures for malaria control in Indonesia: a historical review on species sanitation.* University paper, 90.7.ISBN 90 6754 186 9, Agricultural University, Wageningen, The Netherlands.

Vaas, K.F. (1947) Biologische inventarisatie van den binnenvisserij in Indonesië. *Landbouw (Batavia, Java)*, **29**, 522–543.

van Breemen, M.L. (1930) Verdere gegevens betreffende het malariavraagstuk te Batavia. *Mededeelingen van den Dienst der Volksgezondheid in Nederlands-Indië*, **19**, 197–225.

Walch, E.W., van Breeman, M.L. and Reijntes, E.J. (1930) De assaineering de zoutwater-vischijvers van Batavia, bijdrage tot de ‘hygienische’ exploitatie de bandeng-vijvers. *Mededeelingen van den Dienst der Volksgezonheid in Nederlands-Indië*, **19**, 547–580.

Chapter 14

Socioeconomic and health impacts of water resources development in Thailand

Santasiri Sornmani and Kamolnetr Okanurak

14.1 Introduction

Water resources development (WRD) is important for the strengthening of national economic status. Dams, reservoirs and irrigation systems have been designed and constructed with the primary intentions of converting water into energy and irrigating agricultural land. WRD projects produce relatively clean energy that is cheap for investment and operation. It has been an ideal means of economic development for most countries in the world, particularly during the early period of the twentieth century when many dams and irrigation schemes were built. Thailand has been no exception (Figure 14.1).

The benefits gained from WRD are appreciated by several sectors. For example, urban dwellers enjoy the availability of electricity; industries increase their productivity and create more jobs for workers; farmers cultivate double or more crops resulting in increased agricultural production. While offering these benefits, WRD projects can also have a downside. Several WRD projects, particularly in the tropical countries, have created socioeconomic and health problems for people in project areas. Indigenous peoples lost ancestral lands and their livelihoods in the construction of the Aswan Dam in Egypt (Heyneman 1971; Farid 1975), and at the Bhumiphol Dam project in northern Thailand and the Pattani Dam project in southern Thailand many families had to be evacuated from their homelands before the impoundment (Sornmani and Harinasuta 1988). It seems that in any development project there are always some people who benefit and some people who are disadvantaged. The gains and losses have to be taken into consideration and fairly justified prior to implementation of the project. Mitigation plans should be developed and carried out seriously so that any negative impacts can be avoided, or at least reduced.

The impact of WRD on people will be small or large depending on the type of project, the magnitude of the environmental changes, human behaviour and their interaction with the environment. Generally, multipurpose projects are large-scale and have considerable impact on the environment. For example, a large dam will create a huge man-made lake after the impoundment and the waterbeds in the irrigation area will be full with water all year round. This development and

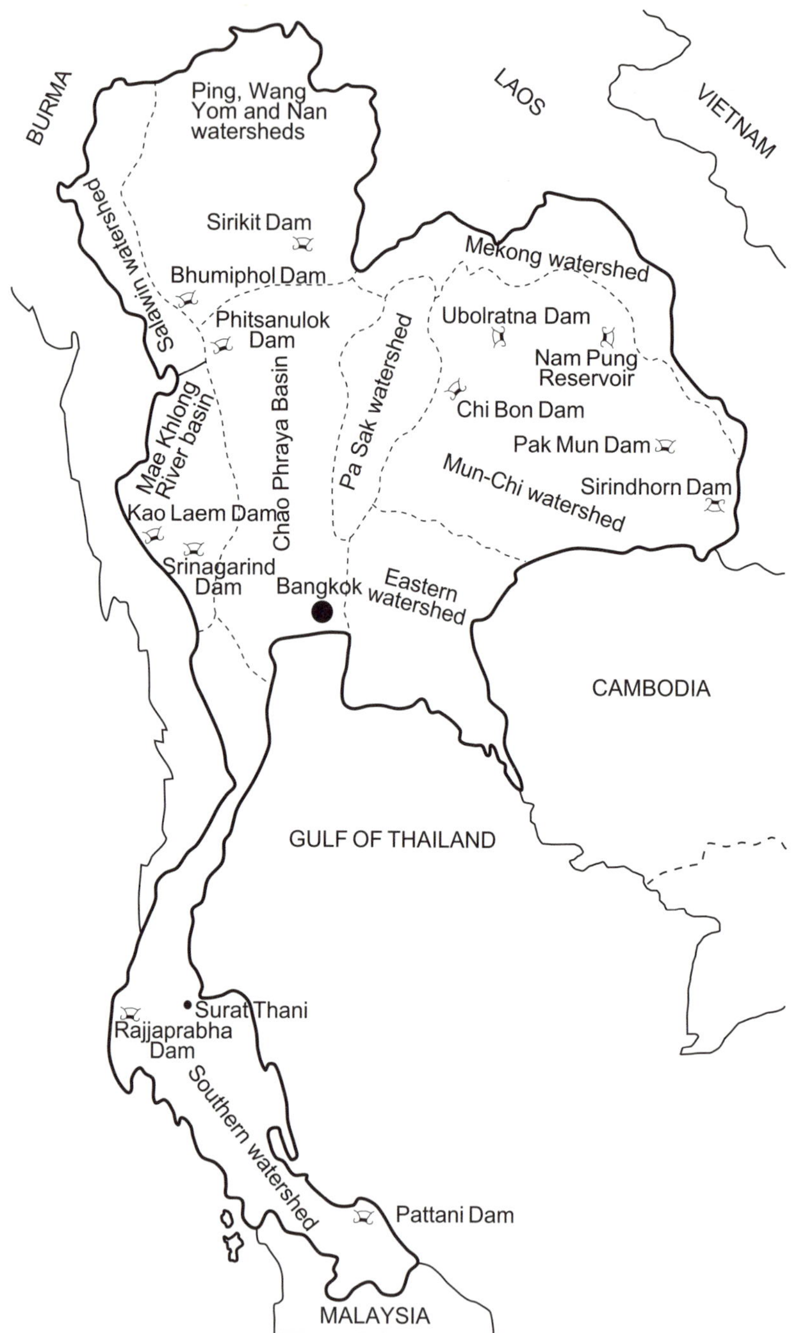

Figure 14.1 The main watersheds and multipurpose dams in Thailand.

environmental change will eventually affect the people connected with the project. Generally in Thailand there are four groups of people that relate somehow to WRD (Sornmani and Harinasuta 1988).

- The people in the inundated area: villagers who reside in the area to be inundated will have to be evacuated before the impoundment. These unfortunate people are forced to migrate from their ancestral land and to resettle elsewhere, usually in a place with unfertile land.
- Labour migrants: this is the pioneer group that migrates into the project, especially to the dam site, during the construction period. The number of migrants varies according to the size of the project. They usually come *en masse*, reside in labour camps and stay for a relatively short period of time. In any project, the labourers can number between 1,000 and 3,000; together with their families the total population can reach 10,000. They may bring infection or be non-immune to infections in the new locality.
- The lake habitants: this group of people comes to settle in the vicinity of the man-made lake after the impoundment. Usually, they arrive in small groups of mobile fishermen one to two years after the impoundment and live on boats, rafts or in huts. As the lake becomes more fertile, more fishermen will come and eventually settle in fishing villages in the drawdown area of the lake. For example, in the Nam Pong project (Ubolratna Dam), there are more than 50 permanent villages with a population of about 30,000. These people earn their living from fishing or culturing fish in the lake.
- The people in the irrigation area: this is the target group for the rural development programme. They are mostly rice farmers who, before construction of the dam, lived in relatively fertile lands. After the construction of the dam and the irrigation system, the government will supply them with modern cultivating technology, improved quality seeds and all year round water. Normally after a few years of working irrigated crops, their incomes increase considerably.

14.2 Social and economic impact of water resources development in Thailand

WRD projects in Thailand have made both positive and negative impacts on the people related to the projects. Evidence from some studies in Thailand demonstrating the socioeconomic impact of WRD on Thai people is presented below.

14.2.1 Economic benefit

Each WRD project has been beneficial to the rural people by increasing family income through increased agricultural production and fisheries. For example, a few years after impoundment in the Nam Pong project, the average annual gross

Table 14.1 Average annual gross income per household in Nam Pong WRD Project in 1973 and the nutritional status of pre-school children (US$ conversions at 1979 rates)

Village	*Sample size*	*Income* *Baht*	*US$*	*% Malnutritional status (weight for age)*
Lakeside	526	15,060	753	10.5
Irrigation	196	20,149	1,009	14.4
Resettlement	63	10,472	524	13.2
Traditional	186	7,969	398	10.0

Source: Suetrong (1979) and Migasena (1978).

income per household in the irrigation area was about Thai baht (THB) 20,150, on the lakeside THB15,060 and among the resettlers THB10,472. Suetrong (1979) showed that people outside the project could only earn about THB7,970 per year (Table 14.1). For comparative purposes, in 1979 the exchange rate was approximately THB20 = US$1; in 1998 it is over THB30 = US$1. The increased earning capacity of people in the project, from either agriculture or fishing, has always been claimed, along with hydropower production, as the economic benefit in every WRD project. However, economic improvement does not always mean an improvement in the quality of life.

14.2.2 Quality of life

It would seem from first principles that if there is more food production from irrigated land and from fish-raising in the man-made lake, then the quality of life in the project area should be better. However, in reality the situation at the Nam Pong project remained the same as before or even deteriorated (Sornmani *et al.* 1973). From data collected by Migasena (1978) 10 years after completion, people in the irrigation area enjoyed a higher income but their children suffered lower nutritional status than children from other areas (Table 14.1).

In the same year, another anthropometric assessment, interpreting wasting and stunting, was conducted in children under five years old residing in the irrigation areas, on the lake shore, in the resettlement villages or in traditional villages outside the project area. Wasting (the result of current nutritional deficiency) and stunting (the result of past nutritional deficits) were prevalent in the children in all villages in the project area (Table 14.2). The nutritional deficit status of children in the project area was more or less similar to that of those in traditional villages outside the project (Schelp *et al.* 1982). This showed that the quality of life of the people in the WRD project did not improve much, even after 10 years. Regardless of the economic gains, the development of children was still as poor as before the project. It is possible that people in the project, especially those in the irrigation areas, were so busy with horticulture and agriculture that they had no time to look after their children.

Table 14.2 Nutritional status of 12–60-month-old children from the Nam Pong project area, lakeshore, resettlement area and traditional villages according to weight for height and height for age (after Schelp *et al.* 1982)

	Normal[a]		*Stunting*[b]		*Wasting*[c]		*Wasting and stunting*[d]	
Area	*No.*	*%*	*No.*	*%*	*No.*	*%*	*No.*	*%*
Irrigation area	491	78.6	44	7.0[e]	79	12.6	11	1.8
Lakeshore	306	75.0	59	14.5[e]	36	8.8	7	1.7
Resettlement area	51	67.1	15	19.8[e]	8	10.5	2	2.6
Traditional villages	114	78.6	16	11.1	15	10.3	–	–
Total	962	76.7	134	10.7	138	11.0	20	1.6

a Based on NCHS standard.
b SD score ≤ 3.00 height for age.
c SD score ≤ 2.00 weight for height.
d SD score ≤ 3.00 height for age and ≤ 2.00 weight for height.
e When tested by chi-square test, statistically significant differences between the irrigation area compared with the lake shore and the resettlement are $p < 0.005$.

14.2.3 Loss of land

The ties that bind people to their homes are very tight and influence people's way of life. In many cases, people consider land to be their nourishing mother or as something to which they belong. For centuries and right up to the present day, people have fought or killed each other because of land. Therefore, loss of land is considered as the greatest loss, not only in terms of occupancy but also in the sense of ancestral property and dignity. At the Bhumiphol Dam in northern Thailand, 11,807 people were forced to migrate from their ancestral land when the lake was formed. They had to sacrifice the fertile land where they had lived for generations and resettle in a remote area. Most of them were relocated to Doi Tao, Chiengmai Province, one of the poorest subdistricts in Thailand. Consequently, they earned less and suffered from a lack of public facilities and water for agricultural and domestic uses (Department of Public Welfare 1977).

14.2.4 Occupational change

In many WRD schemes people have to change their occupations, e.g. from rice cultivation to pararubber plantation or fisheries. Because people in developing countries are used to earning their living by agricultural activities, it has been integrated into their way of life and enrooted in their tradition and culture. Therefore, any changes in agricultural activities are associated not only with changed occupational status but also with way of life. If such people are persuaded to grow more crops for the sake of getting a cash income, this change in itself is not essentially disruptive. However, when the change is from subsistence agriculture to cash crops, a number of disruptive factors may enter the picture. A good example is the case of the Rajjaprabha project in southern Thailand. People were advised to change from growing fruit trees and field crops to pararubber plantation. The most immediate problem was that they did not know how to cultivate pararubber. The authority had to organize training courses to educate the resettled population and had to assist them in growing and taking care of rubber trees. Even with such assistance, not all the resettlers liked or wanted to live with the new occupation. Some moved away and preferred to retain their original occupation elsewhere (Team Consulting Engineers and Department of Community Development 1980).

14.2.5 Psychosocial problems

For the large majority of any population, compulsory relocation is a traumatic experience that causes multidimensional psychological stress. Because most people do not wish to move, they resist the attempts of the resettlement administration. However, their resistance is finally broken and movement occurs. Not only are the people uprooted from their homes and occupations, but they also are faced with the question of their ability to live in the new environment.

In the case of the Rajjaprabha (Chiew Larn) project, the first problem was associated with the topography, from flat terrain along the river to an upland area with undulating ground. Second, since pararubber was the recommended primary crop, the cultivation practice changed from fruit tree and field crop farming to pararubber cultivation. Third, community structure was changed from a line type settlement in small sporadic clusters along the river to rows of houses in large clusters. Some families had to live close to the indigenous dwellers of the resettlement site who regarded the resettlers as newcomers and not as neighbours. All of these changes caused psychosocial problems in the new community (Team Consulting Engineers and Department of Community Development 1980).

14.2.6 Loss of archaeological evidence

The history of mankind indicates that in ancient times peoples lived scattered all over the place. This is evident from discoveries of statues, artefacts and skeletons in caves, under rice fields or in sugar plantations. Man-made lakes have submerged these, however. In the case of the Aswan Dam, ancient statues and temples such as Abu Simbel reflecting the glory of the Egyptian era were partly removed to alternative places above water level. This was expensive and involved a risk of damage during removal. In Kanchanaburi Province in central Thailand, thirteen archaeological sites lie under the Srinagarind reservoir (Southeast Asia Technology and National Institute of Development Administration 1987). Although these sites may not be recognized as world heritage, they have a great psychological value to the Thai people.

14.2.7 The readiness of people to make use of water

Physical facilities are always the main core of development. Very often, these facilities are misused or underutilized if people have not been prepared and trained to use them. In the Nong Wai irrigation scheme in Khon Kaen Province, it took a few years for people to fully utilize irrigation water because they had only had experience with rain-fed ecosystems. Despite the availability of water in the irrigation canals to cultivate rice or cash crops during the dry season, the people had decided that the growing of crops in an unfamiliar season was a very difficult job. Irrigation water was used later, but only after the people had been taught how and had got used to new kinds of seeds and new dry season cultivation techniques.

During the first few years after impoundment of water in the Ubolratna dam, old style small and slender boats previously used in the old river were used in this large lake of 410 km^2. Several boats capsized and lives were lost because they had little experience with waves and wind.

14.3 The health impact of water resources development projects in Thailand

Because early schemes were built in temperate areas where water-related parasitic disease was absent, the impact of WRD projects on public health was seldom recognized. Problems gradually became apparent when developments were carried out in tropical countries where water-related diseases were abundant. These diseases, with few exceptions, take time to develop into visible and permanent morbidity in the population. Furthermore, the demonstration of a health impact requires longitudinal study over years, and great expense. For these reasons, in South-East Asia, particularly in Thailand, it took decades for countries to realize the health impacts of WRD. Generally speaking, the major impact on health in Thailand is of two kinds: the problem of vector-borne diseases; and the problem of water quality.

14.3.1 The problem of snail-borne diseases

The important snail-borne diseases are the parasitic trematode or fluke infections, namely schistosomiasis (blood fluke infection), opisthorchiasis (liver fluke infection), paragonimiasis (lung fluke infection) and other intestinal fluke infections, such as echinostomiasis, fasciolopsiasis and heterophysiasis. To complete their life-cycles, the parasites of this group need specific kinds of snails to serve as their intermediate hosts, and then either snail, crab, waterplant, tadpole or dragon fly nymph to serve as their second intermediate host. In most cases, except for schistosomiasis, the second intermediate host harbours the infective stage of the parasite and gains access into the human body via ingestion. Infections of liver fluke, lung fluke and intestinal fluke are commonly found in Thailand where people are fond of eating raw food. Cyprinoid fish, crab, snail and tadpoles are the main sources of these infections.

In addition to trematodes, other nematode or round worm infections, such as *Angiostrongylus cantonensis* (rat lung worm), are also prevalent in certain localities of Thailand. The infection is caused by eating raw *Pila* snails, which harbour the larval stage of this parasite. The infection appears as eosinophilic meningitis and can be fatal in heavy infections.

Opisthorchiasis

Of all the trematode infections, opisthorchiasis has been shown to be the most related to water resource developments (Harinasuta 1975). Because of the availability of water all year round in these projects, populations of *Bithynia* snails and cyprinoid fish become abundant. Because local inhabitants sometimes defecate outside of latrines and eat raw fish, the infection can spread widely. The disease is chronic and the patients do not suffer much in the early stage of the infection. From time to time, the patients will became ill, weak and unable to work. In the late stage, damage to the biliary system and liver is beyond repair. Patients may die from biliary fibrosis of the liver and often from carcinoma of the biliary system.

In the study at the Nam Pong WRD project in Thailand, it was clearly demonstrated that despite the development and increased income, people in the irrigation area were suffering from illnesses similar to those in the undeveloped area. The prevalence of opisthorchiasis among the people in the irrigated area was higher than in the non-irrigated area (Harinasuta *et al.* 1976; Sornmani *et al.* 1981; Sornmani and Harinasuta 1988). Because of the irrigation system, water is available all year round, which offers many more habitats for snails and fish. The chance of spreading opisthorchiasis to other populations is high since the fish caught from the lake and environs are sold all over the region.

From an economic point of view, it has been estimated that in 1981 there were about seven million people infected with opisthorchiasis (Preuksaraj *et al.* 1982). The approximate economic loss has been calculated at US$19.4 million for treatment and US$65 million for wage losses (Loaharanu and Sornmani 1991).

Schistosomiasis

The implication of schistosomiasis in WRD projects in Thailand needs close and longitudinal observation. There were a few reported cases that were accidentally found in some patients who had attended hospitals for unrelated illnesses. Most of these cases were traced back to their villages and epidemiological surveys were made in these areas, but no other active infections have ever been detected. Snail surveys were also conducted, but none was found to be infected. In the pre-impoundment survey for snail-borne diseases in the Rajjaprabha Dam project,

Surat Thani Province in southern Thailand, live *Schistosoma japonicum*-like eggs were found in the stools of a boy and his dog but, as before, further surveys were negative (Sornmani 1988).

In all water resource development projects in Thailand, the problem of schistosomiasis is not obvious, but it is prudent to set up surveillance systems in some projects. The important one is the Pak Mun project on the Mun River in Ubol Ratchathani Province in north-eastern Thailand. This is a good model for the study of the social and health implications of water resource development in Thailand and other riparian countries along the Mekong River. This will be detailed later.

14.3.2 The problem of mosquito-borne diseases

The mosquito-borne diseases that are considered to be related to WRD projects in Thailand are malaria, filariasis, dengue haemorrhagic fever (DHF) and Japanese encephalitis (JE).

Malaria and filariasis

These diseases cause major problems in water resource development projects in South-East Asia. However, these diseases are mostly only a problem during the construction phase and when the project is in the jungle. Since most multipurpose projects are large-scale and need a large amount of labour, migration of skilled or unskilled people is always required. The chance of exposing these non-immune people to malaria is high, and the risk greater than that for the local indigenous population. In the construction of Rajjaprabha Dam, more Thai engineers and migrant labourers suffered from malaria than did indigenous people.

Resettlement poses another problem. In multipurpose projects, evacuees from the flooded area are often resettled in unfertile land on the forest fringe where they are bitten by *Anopheles minimus* and *Anopheles dirus*, both forest dwellers and major malaria vectors. Malaria was shown to be a major health problem in the resettlement villages of the Srinagarind and Rajjaprabha Dam projects in Thailand. Living along the forest fringe, the settlers were faced with jungle breeding mosquito *Anopheles dirus* (Sornmani 1988). Malaria control measures, such as active case detection and annual DDT residual spraying had to be implemented, otherwise people could not reside in those resettlement villages.

The relationship between filariasis and water resources development projects has not been elucidated. The spread of filariasis to the labour force during the construction phase of projects is likely to occur, but is not yet evident in South-East Asia. It is well known that the lag between infection with microfilaria from mosquito bites and the occurrence of symptoms may be 15 years. Because the Srinagarind Dam project is in an area endemic for bancroftian filariasis, periodic monitoring is necessary.

Arboviruses

DHF is carried by *Aedes* mosquitoes that breed in artificial containers. Because resettlements or lakeside villages may not have piped water, water is often stored in giant water jars, in which the mosquitoes breed, which the inhabitants can only afford to buy because their income has increased. This provides the chance for DHF to spread via increased numbers of *Aedes aegypti*.

The mosquito vectors of JE, on the other hand, breed in rice fields. The extra irrigated water facilitates double or triple cropping, thus providing the chance for JE vectors to breed more. At Srinagarind, *Aedes albopictus* (dengue) and *Culex tritaeniorhynchus* were breeding at the reservoir (Gass 1988; Sornmani 1989). So far, there is no evidence to incriminate the reservoir as a source of JE infection but this possibility has to be monitored.

14.3.3 The problem of poor water quality

Only recently has the impact of poor water quality from water resource projects become evident in Thailand. The problem usually occurs only in the deep reservoirs and affects downstream populations. Such reservoirs have large volumes of unoxygenated water that is collected and released through turbines several kilometres downstream. Rajjaprabha Dam used to release water with low oxygen content downstream past several villages located on both banks of the Khong Saeng River. The water quality was so poor that the people could not use it for bathing or washing. Complaints were made to the government, and finally huge air pumps were installed to increase the oxygen content of the water prior to release. However, this provided only a temporary solution and eventually the project management had to provide every village with an artesian well.

A similar situation arose at the Srinagarind Dam but the impact was not as serious because there were no people living immediately below the dam. By the time the water reached the villages, water quality had been gradually improved by the flow. The quality of discharged water is not usually included in health impact considerations, but it should be taken very seriously, particularly if there are people living downstream.

14.4 Pak Mun project

The Mun River is the most important river in the north-eastern region of Thailand (Figure 14.2). It is in the Mun–Chi watershed, which has a total drainage area of 117,000 km^2. Influenced by the tropical monsoons, the average annual rainfall over the basin is 1,307.4 mm. The main stream of Mun River is in Ubol Ratchathani Province and drains into the Mekong River, with an average annual run-off about 20,381 million m^3.

The Pak Mun Dam site is on the Mun River, 82.5 km to the east and downstream of the town of Ubol Ratchathani or about 6 km upstream from its

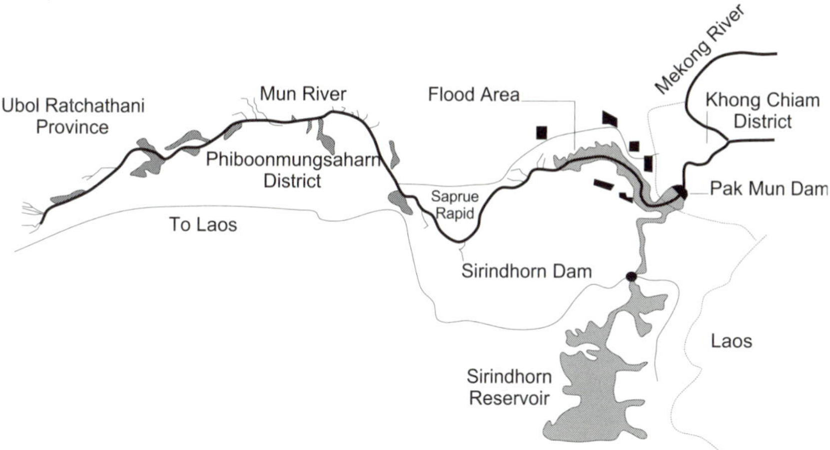

Figure 14.2 The Pak Mun project and flood area in the Mun River and Sirindhorn Dam.

confluence with the Mekong River. The dam is roller compacted concrete with a height of 17 m and a length of 300 m. The normal water level of the river upstream is 108 m above mean sea level (MSL). As a river-run type, the Pak Mun Dam was designed to raise the water level of the Mun River to the full bank level. The storage will utilize only the capacity of the river channel, which has a surface area of 60 km^2 and a storage area of 225 million m^3. The generating units of Pak Mun power plant are bulb turbine generators which were designed to operate with 3–14 m head of water. The power plant can produce 280 million kWh annually. This energy will strengthen the stability of the north-eastern region power supply system. The project will be able to irrigate 26,000 ha of land on both sides of the Mun River. The upstream river reservoir is claimed to be suitable for fisheries development of about 1.3 million kg per year. The construction took four and a half years to complete at a cost of THB6,600 million (approximately US$264 million at 1993 exchange rates). In September 1994, the construction was completed and the generators started to produce electricity.

There were thirty-one villages or 989 households affected by the flooding of 912 ha. They were either in the flood area (below 108 m MSL), or living at the dam site. Nearly THB 200 million (US$8 million at 1993 rates) was paid in compensation for loss of land, residences, trees and other properties and for the construction of the resettlement site. There are still some compensation claims under negotiation between the villagers and project management.

14.4.1 The socioeconomic problems

The Pak Mun Dam is relatively small in size when compared with other multi-purpose projects, such as the Bhumiphol or Pattani Dams. However, there were a lot of problems because the project was constructed in a populated area. The other projects were in forest with few indigenous people. The major concerns of the people affected directly or indirectly by the Pak Mun Project centred on whether they had to be evacuated from their homeland, and whether the reservoir would reduce the fish population and, accordingly, their incomes.

Although the project management had defined the area and population to be affected and those who would receive compensation, there were some people who did not want to move out from their home and settle elsewhere. This resulted in many complaints about compensation and resettlement areas. This is acceptable and expected to occur in any WRD. It was resolved by negotiation, increased compensation and, in a few cases, by force.

In addition, there were some people who did not know the meaning of 108 m MSL or did not believe that the water level marked by the project survey team was correct. Arguments occurred and the government had to send an independent team to measure and mark the maximal water level again and again. Even when the level was agreed upon, there were still more than 670 villagers who were afraid that the flood would be over the 108 m MSL level (Electricity Generating Authority of Thailand 1993). These people were not included in the early compensation group, but they asked for the right to be compensated too. This problem is still under negotiation.

A section of the downstream river bed was dynamited to allow free flow of the water from the dam. Occasionally rock debris reached some houses on the banks of the river. This activity became a big issue, raised by 238 villagers and the protesters against dam construction. The argument was that the rock blasting was not made known to the public before the construction and it endangered their lives and properties. The project was forced to stop working until agreement was reached by negotiation. Some households in this village were relocated to other places and paid extra compensation.

The Mun River is a tributary of the Mekong River, and fish usually migrate upstream to increase the fish populations and generally maintain the fauna in the river. To mitigate the potential fishing problem, the project management provided THB 16 million (US$ 0.64 million) at the design stage to implement fish seeding, a fish ladder and a fisheries resources conservation programme. Nevertheless, people remained concerned. It is still not yet certain whether this fisheries mitigation plan has worked.

In addition to the problem of fish populations, there is another problem that has dragged on over the past three years. In negotiations for compensation, the project management was asked to pay compensation to fishermen who would lose their occupation or would earn less after the impoundment. Along this river, there were more than 200 villages where many people fish; some took fishing as their main occupation, while others might do it occasionally. It became difficult to separate the real fishermen from the occasional fishermen when all of them claimed that they were professionals. Several protests were made and several committees were appointed to justify compensation. Negotiations are still going on.

These above mentioned problems comprise the socioeconomic impact originating from the Pak Mun project. Some of these problems are not new since they have also been seen in former projects in Thailand. However, these problems have been intensified because the project is located in a populated area. As such, this debate has been greater than that over any other Thai project. It is difficult to justify who is right and who is wrong. It depends on what angle one is looking at. Nevertheless, one has to keep in mind that in any development project there will be some advantages and some disadvantages. The disadvantage in social terms is difficult to evaluate in monetary terms, but the disagreement, protest and non-cooperation of the people in the project area can severely interfere with its progress. On several occasions, construction at the Pak Mun had to be stopped until the social problems had been cleared up. The total cost to the project, not counting the time lost, was more than 50 per cent of the original budget.

14.4.2 Health impact of the Pak Mun project

Because the Pak Mun project was proposed for development more than a decade ago, several studies have been made in the area. Based on these the major concerns about the impact of the Pak Mun project on the health of the people in the area can be considered as two problems: the spread of the local endemic disease opisthorchiasis and the possible establishment of *Schistosoma mekongi* in the Mun River.

Opisthorchiasis

In the Pak Mun project, infected snail hosts and infected fish are plentiful in the area and so the opportunity for the spread of infection exists even without the construction of the dam. However, there are two factors that may mitigate transmission. The National Liver Fluke Control Programme was launched in 1990. Everyone under this programme has an annual stool examination for the *Opisthorchis viverrini* eggs and, if positive, is treated with the anthelmintic drug praziquantel. The programme is organized by the Ministry of Public Health and is implemented locally by the health officers in every subdistrict of the region. The programme is supported by the Liver Fluke Control Programme, which not only does the stool examinations and treatments but also provides health education aimed at improving the eating and defecating habits of the people. Since the programme has been in continuous operation for sometime, the overall annual prevalence rate has decreased.

The latest parasitic survey made by the Ministry of Public Health in Mun River basin has also shown a decreasing trend of liver fluke infection (Jongsuksuntigul *et al.* 1992). *Bithynia* snails usually live in stagnant water beds, such as rice fields, canals and pools and are very seldom found in the rivers. As the Pak Mun Dam is a river reservoir, it does not create stagnant water. Nevertheless, a surveillance programme for liver fluke infection has to be set up and if there is a trend of increasing prevalence, control measures should be actively implemented.

Schistosomiasis

Besides liver fluke infections, *Schistosomia mekongi* is an important parasite that needs to be considered in the Pak Mun project. *Schistosomia mekongi* is endemic at Khong Island, southern Laos. This island is about 200 km downstream in the Mekong River from the Pak Mun Dam (Sornmani *et al.* 1971). *Neotricula aperta* has been identified as the snail intermediate host of this parasite. These snails comprise three races: alpha, beta and gamma (Davis *et al.* 1976). Alpha and gamma races were found in the Mekong River from the Khemmarat District to Khong Chiam in Thailand and at Khong Island in Laos. *Neotricula aperta*, beta race, was the only race found in Mun River in Thailand between the town of Ubol Ratchathani and its confluence to the Mekong River at Khong Chiam District (Sornmani 1976a). This part of the river is the reservoir of the Pak Mun Dam. The beta race *Neotricula* snails normally live in running water, such as on rocks in rapids or on rock islands in the middle of the Mun River. They are rare in rock pools because they do not live long under these conditions. Snails colonize by laying eggs on the undersurface of rocks, dead wood and sometimes on the stems of waterplants. The running water in their habitats must be clean with no mud or algae. The eggs hatch in early summer (March–April) when the water level is low, and the baby snails develop in six to eight weeks into adult snails using diatoms on the rock surface as their food. In the beginning of the rainy season (June–July) or when the water level starts rising, they lay their eggs on the rock surface. In late rainy season (November–December), the water level in the river is several metres higher than in summer. At this time, no snails are found in these habitats, as they either die or are washed away by the strong current (Upatham *et al.* 1980). No natural schistosome infection has ever been found in the beta race *Neotricula*. The only natural infection has been found in the gamma race *Neotricula* at Khong Island (Sornmani 1976b).

A recent survey (Sornmani *et al.* 1994) for *Neotricula* snails in the reservoir of the Sirindhorn Dam in the Mun River Basin and close to the Pak Mun Dam proved negative although this reservoir is 20 years old. In Ubol Ratchathani hospital in 1966, *Schistosoma* infection was accidentally found in an autopsied specimen. A follow-up survey detected a Laotian patient in the hospital and a woman in a village near the Thai–Laos border infected with schistosomiasis, but the place of transmission was not known (Desowitz *et al.* 1967). Epidemiological surveys by the Faculty of Tropical Medicine, Mahidol University, Bangkok have been carried out from time to time, but no schistosome infection was ever found in humans or animals in this area. In addition, the surveys on water-borne diseases in the environmental and ecological impact assessment of Pak Mun Dam in 1981 and 1984 (Team Consulting Engineers and the Faculty of Tropical Medicine 1984) and the latest survey for surveillance of water-borne parasitic diseases in the Mun River basin (Jongsuksuntigul *et al.* 1992) were also negative.

It is likely that conditions will not favour the *Schistosoma–Neotricula* relationship in the river and that the foci of transmission will be difficult to establish. According to the engineering concept, the Pak Mun Dam will regulate water levels in the river so that they are similar to the natural conditions of the Mun River before damming. Therefore, the chance of *Schistosoma* infection becoming established in the river after dam construction would be more or less the same as that without construction.

In any case, the schistosomiasis situation in the Mun River basin has to be monitored. From this point of view, the Pak Mun project has already included in its health mitigation plan a monitoring system for schistosomiasis in the human population in the project area and of *Neotricula* snails in the river. The first baseline survey was completed in 1993 and was negative.

In conclusion, the Pak Mun project provides an interesting model for learning about the socioeconomic and health impact of water resources development in the Lower Mekong basin, particularly in southern Laos and northern Cambodia. The problems described above are distinctive and seldom discussed. Some of them were not anticipated and some have expanded into political issues. At present, our knowledge about the biology and biopathology of *Neotricula* snails and *Schistosoma mekongi* is relatively limited in comparison with that for other human *Schistosoma* infections in other parts of the world. Therefore, in some ways it would seem rather ambitious to forecast the health impact of water resources development in the Lower Mekong basin with an inadequate data base. It is necessary that research in this field should be encouraged so that results will be used to improve the quality of life of the people in this part of the world.

References

Davis, G.M., Kittikoon, V. and Temcharoen, P. (1976) Monograph on *Lithoglyphosis aperta*, the snail host of Mekong schistosomiasis. *Malacologia*, **15**, 241.

Department of Public Welfare (1977) *Resettlement Area of Bhumipol Dam, Chiengmai Province* (in Thai). Ministry of the Interior, Bangkok.

Desowitz, R., Harinasuta, C., Krustrachue, M., Chesdaphan, C. and Jetanasen, S. (1967) The result of a stool and skin test survey for schistosomiasis in villages near Mekong river of northeast Thailand. *Transactions of the Royal Society for Tropical Medicine and Hygiene*, **61**, 153.

Electricity Generating Authority of Thailand (1993) *Report on Pak Mun Hydroelectric Project for IBRD Missions, Bangkok*. Electricity Generating Authority, Bangkok.

Farid, M.A. (1975) The Aswan High Dam Development Project. In: *Man-made Lakes and Human Health* (eds Stanley, N.F. and Alpers, M.P.). Academic Press, London, pp. 89–102.

Gass, R.F. (1988) *A Longitudinal Mosquito Survey at the Srinakarind Reservoir: An Assessment of the Changes in Abundance and Behavior of Vector Species Induced by Dam-associated Ecological Disturbances*. Faculty of Tropical Medicine, Mahidol University, Bangkok.

Harinasuta, C. (1975) Ubolratana Dam Complex, Thailand. In: *Man-made Lakes and Human Health* (eds Stanley, N. P. and Alpers, M.P.). Academic Press, London, pp. 149–164.

Harinasuta, C., Sornmani, S., Migasena, P., Vivatanasesth, P., Pongpaew, P., Intrakhao, C. and Vudhivai, N. (1976) Socio-economic, health and nutritional status of villagers in the Nong Wai Irrigation Area, Khon Kaen, Northeast Thailand. *Southeast Asian Journal of Tropical Medicine and Public Health*, **7**, 601.

Heyneman, D. (1971) Mis-aid to the Third World: disease repercussions caused by ecological ignorance. *Canadian Journal of Public Health*, **62**, 303.

Jongsuksuntigul, P., Chaychumsri, W., Techamontrikul, P., Jeradit, P. and Suratawanich, P. (1992) *Studies on Prevalence and Intensity of Intestinal Helminthiasis and Liver Fluke In Thailand 1991*. Report of the National Survey by the Health Division, Ministry of Public Health, Bangkok.

Loaharanu, P. and Sornmani, S. (1991) Preliminary estimates of economic impact of liver fluke infection in Thailand and the feasibility of irradiation as a control measure. *Southeast Asian Journal of Tropical Medicine Public Health*, **22**, 384.

Migasena, P. (1978) Public health and nutrition. In: *Study of Environmental Impact of Nam Pong Project, Northeast Thailand*. SEATEC Report prepared for National Energy Administration, Office of the Prime Minister, Bangkok.

Preuksaraj, S., Jeradit, C., Sathitayathai, A., Kijvannee, S. and Seedonrusmi, T. (1982) Studies on prevalence and intensity of intestinal helminths in the rural population of Thailand 1980–1981. *Communicable Diseases Journal*, **8**, 245 (in Thai).

Schelp, F.P., Sornmani, S., Pongpaew, P., Vudhivai, N., Egormaiphol, S. and Bohning, D. (1982) Seasonal variation of wasting and stunting in preschool children during a three-year community-based nutritional intervention study in Northeast Thailand. *Tropical Medicine and Parasitology*, **41**, 279–285.

Sornmani, S. (1976a) Current status of schistosomiasis in Laos, Thailand and Malaysia. *Southeast Asian Journal of Tropical Medicine and Public Health*, **7**, 149.

Sornmani, S. (1976b) Current status of research on the biology of Mekong *Schistosoma*. *Southeast Asian Journal of Tropical Medicine and Public Health*, **7**, 208.

Sornmani, S. (1988) *The Study of Schistosomiasis in the Rajjaprabha Project*. Report to Electricity Authority of Thailand, Bangkok University Faculty of Tropical Medicine, Mahidol University, Bangkok.

Sornmani, S. (1989) Impact of water resources development on disease vectors. In: *Development of the Health of Communities and Preventive Measures for Adverse Effect* (eds Bunnag, T. and Sornmani, S). 13–16 June 1988, Surat Thani, Thailand, pp. 120–131.

Sornmani, S. and Harinasuta, C. (1988) *Water Resources Development and Its Impact on Socio-Economic and Health with Reference to Thailand*. Information Centre, Faculty of Tropical Medicine, Mahidol University, Bangkok.

Sornmani, S., Kitikoon, V., Harinasuta, C., and Pathammavong, O. (1971) Epidemiological study of *Schistosoma japonica* on Khong Island, Southern Laos. *Southeast Asian Journal of Tropical Medicine and Public Health*, **2**, 365.

Sornmani, S., Schelp, F.P. and Harinasuta, C. (1981) Health and nutritional problems in the Nam Pong Water Resource Development Scheme. *Southeast Asian Journal of Tropical Medicine and Public Health*, **12**, 402.

Sornmani, S., Suthimonthon, C., Okanurak, K. *et al.* (1994) *Investigation on the opportunity of* Neotricula aperta *snail intermediate host of* Schistosoma mekongi *to colonize in the reservoir and the survey for Schistosomiasis in the vicinity of the reservoir of Sirindhorn Dam, Ubol Ratchathani Province*. Report to Electricity Authority of Thailand, Bangkok, Faculty of Tropical Medicine, Mahidol University, Bangkok.

Sornmani, S., Vivatanasesth, P., Bunnag, T., Intrakhao and Harinasuta, C. (1973) A study on the pattern of socio-economic and health status in relation to parasitic diseases in the inhabitants around Ubol Ratana Dam in northeast Thailand. *Southeast Asian Journal of Tropical Medicine and Public Health*, **4**, 421.

Southeast Asia Technology and National Institute of Development Administration (1987) *Post Environmental Evaluation of Srinagarind Dam*. Vol. II (Part Four), Bangkok.

Suetrong, S. (1979) *Nam Pong Environmental Management Research Project*. Working document No. 8, Socio-economic Studies. Mekong Committee, Bangkok.

Team Consulting Engineers and Department of Community Development (1980) *Chiew Larn Project Environmental and Ecological Investigation*, Final Report Vol. II, Faculty of Social Administration, Thammasart U. Main Report, Electricity Generating Authority of Thailand, Bangkok.

Team Consulting Engineers and Faculty of Tropical Medicine (1984) *Final Report on Selected Environmental and Ecological Investigation of Pak Mun Project*, Vol. I: Summary of Environmental and Ecological Impact Assessment. Electricity Generating Authority of Thailand, Bangkok.

Upatham, E.S., Sornmani, S., Thirachantra, S. and Sitaputra, P. (1980) Field studies on the bionomics of alpha and gamma races of *Tricula aperta* in the Mekong River at Khemmarat, Ubol Ratchathani Province, Thailand. In: *The Mekong Schistosome* (eds Bruce, J. I. and Sornmani, S.), *Malacological Review*, Supplement 2, p. 263.

Chapter 15

Urban water resource problems in vector-borne disease with special reference to dengue viruses

Brian Kay

15.1 Problem definition

According to 1991 data from the Organization for Economic Co-operation and Development (1991), domestic and municipal water accounts for 7 per cent of total withdrawals although usage is substantially higher in developed countries. The energy and industrial sectors in developed countries account for 40–80 per cent, whereas it is only a few per cent in developing countries. Given a typical cost of US$120 per person served by household reticulation in developing countries (Cairncross 1990), a cheaper alternative is the public stand-pipe. Savings not only occur with respect to lowered costs for the distribution system itself, but also mainly because households relying on public access consume far less water than do those with private connections.

Figure 15.1 Water storage Brazil style. Donkey power is used to deliver water drawn from the public cistern to householders.

Lack of a reliable piped water system therefore reduces the amount of water available for washing, cooking and other domestic activities. It confers a greater risk of contracting water-borne or water-washed infections causing enteric disease. The other problem is that water storage in wells, jars, tanks, cisterns or drums becomes obligatory and these provide ready-made breeding habitats for mosquitoes to cause water-related disease (Figure 15.1). In some situations, although piped water systems have been installed, communities will not necessarily remove or destroy their water storage containers because of their cultural upbringing and/or possibly because of innate distrust that such a system will be reliable.

Of particular concern and recognized by the World Health Assembly (1993) as a priority are the dengue viruses, which are mainly transmitted by *Aedes aegypti* and other *Stegomyia* group mosquitoes. Since 1953, dengue haemorrhagic fever has emerged in forty-six countries, with approximately two-fifths of the global population from ninety-four countries at risk (Table 15.1). Dengue haemorrhagic fever, which causes leakage of primarily the white cells out of the vascular system, can be fatal to children. Epidemics can sweep through urban and rural communities within a few months.

Urban malaria occurs at approximately 50,000 cases per year in cities such as Madras. The primary cause of this are *Anopheles stephensi* mosquitoes, which breed in roof-top cisterns. Most sanitary systems require far less water than conventional sewage systems and sewage treatment plants. However, at US$75–150, the average cost of a twin-pit pour-flush latrine represents the whole annual income of some people, and a conventional septic tank, at US$200–600 per household (Sinnatamby 1990), would seem totally out of the question. Unfortunately, primitive latrines form a haven for mosquitoes such as *Culex quinquefasciatus*, which prefer polluted waters. This, in turn, creates problems with lymphatic filariasis caused by the parasitic worm *Wuchereria bancrofti*.

The provision of adequate drainage to remove storm water and domestic and industrial effluent is important in reducing mosquito breeding. Gully traps and sumps are often designed with silt traps so that the outlet pipe is higher than the base. Unfortunately, water often collects in these wastes and mosquitoes then breed there. Cairncross and Feachem (1983) cover health engineering in the tropics. Moist soil around stand-pipes can also harbour parasites such as *Ascaris* and hookworm. The problem is therefore very much an intersectoral one and one in which health authorities should strive to develop workable relations with planners, engineers and developers, especially at the municipal level. The lack of an adequate solid waste management system not only increases the risk of polio, enteric disease and neonatal tetanus, but also creates a multitude of small artificial breeding sites such as tins, tyres, bottles, metal parts, etc. that are suitable for dengue mosquitoes. Just as it is important to consider both input and output with respect to provision of adequate drainage systems, it is also important that, if building regulations have been legislated, then the resulting structures must be inspected on completion. All too often, building applications pass through several

Table 15.1 Countries or territories in which dengue or dengue haemorrhagic fever is known to occur, by WHO region, 1975–1993

Africa (18)		
Burkina Faso	Kenya	Sierra Leone
Comoros	Madagascar[a]	Somalia
Ivory Coast	Mozambique	South Africa[a]
Ethiopia	Nigeria	Sudan
Ghana	Senegal	Tanzania
Guinea	Seychelles	Zaire
The Americas (40)		
Antigua and Barbuda	Ecuador	Paraguay
Aruba	El Salvador	Puerto Rico
Bahamas	French Guiana	St Kitts and Nevis
Barbados	Grenada	St Lucia
Belize	Guadeloupe	St Martin
Bolivia	Guatemala	St Vincent and the Grenadines
Bonaire	Guyana	Suriname
Brazil	Haiti	Trinidad and Tobago
British Virgin Islands	Honduras	Turks and Caicos Islands
Curaçao	Jamaica	USA[b]
Colombia	Martinique	Venezuela
Cuba	Mexico	US Virgin Islands
Dominica	Montserrat	
Dominican Republic	Nicaragua	
South-East Asia (7)		
Bangladesh	Maldives	Thailand
India	Myanmar	
Indonesia	Sri Lanka	
Europe (1)		
Bosnia and Herzegovina		
Eastern Mediterranean		
(No cases reported)		
Western Pacific (28)		
American Samoa	Malaysia	Solomon Islands
Australia	Marshall Islands	Tokelau Islands
Cambodia	Nauru	Tonga
China	New Caledonia	Tuvalu
Cook Islands	New Zealand	Vanuatu
Fiji	Niue	Vietnam
French Polynesia	Palau	Wallis and Futuna Islands
Guam	Papua New Guinea	Western Samoa
Kiribati	Philippines	
Laos	Singapore	

[a] Possibly imported.
[b] Imported.
Source: World Health Assembly (1993).

government departments during the permitting process but after that the builders can do what they wish.

Inadequate storm water pipes and drainage systems can lead to seepages in low-lying areas; roof gutters can become blocked leading to unrecognized mosquito colonizations. The practice of using polyvinyl chloride (PVC) pipes over land-fills sometimes results in bowed pipes and undetected water storage. At Charters Towers in northern Australia, the dengue mosquito *Aedes aegypti* and other tree-hole breeders readily colonize water stored in PVC piping linking houses to kerbside guttering. Such subterranean habitats provide ideal conditions for mosquitoes all year round. Subterranean breeding was linked to the 1992–93 outbreak of dengue fever in Charters Towers.

The use of wells and septic tanks as mosquito breeding sites has been documented, particularly in India, Sudan and the Middle East. In India and Iran, well-breeding *Anopheles* have been linked to malaria transmission (Batra and Reuben 1979; Eshghy 1977). Wells are also habitat for *Culex quinquefasciatus*, e.g. in India (Yasuno *et al.* 1977) or Brazil (Kay *et al.* 1992), and *Aedes aegypti*, e.g. in India (Panicker *et al.* 1982), Laos (Jennings *et al.* 1995) or Australia (Russell *et al.* 1996). Other subterranean habitats are created in the name of development. Underground pumping stations may leak and form *Aedes aegypti* habitat as they did in Townsville during 1992–93. Precast concrete service manholes used for inspection of telecommunications and other cables may contain over 1200 l of water, and with it large populations of mosquito larvae. In northern Queensland, subterranean habitats provide refuge from adverse conditions during winter or the dry season. The importance of such sites is underrated, and because they often are difficult to detect and inspect they are often neglected in public health programmes.

The core of the problem, of course, is burgeoning urbanization and the understandable difficulties that confront municipal and government authorities in providing an adequate infrastructure and services to cater for the influx. Cities are currently absorbing two-thirds of the total population increase in developing nations. During the 1950s, only 29 per cent lived in urban areas; between 1960 and 1980, urban populations doubled in developing countries (Knudsen 1992). Based on 1992 revisions 50 per cent and 60 per cent of the global population will live in cities in 2005 and 2025 (United Nations 1993). In 1970, five of ten cities with over eight million residents (megalopolis) were in the developing world, but by 1990 this had extended to fourteen out of twenty. By 2010, it is expected that twenty-four out of thirty will be in developing countries and sixteen in Asia (Table 15.2). Only a few of these will be spared from explosive outbreaks of dengue and dengue haemorrhagic fever. At present, dengue and dengue haemorrhagic fever threaten two-fifths of the global population in ninety-four countries in Africa, the Americas, Asia, Australia and the Pacific Islands. What will it be like in 2010?

Table 15.2 Urban agglomerations of eight million or more people, by development regions

More developed regions				*Less developed regions*			
1950	*1970*	*1990*	*2010*	*1950*	*1970*	*1990*	*2010*
New York	Tokyo	Tokyo	Tokyo	None	Shanghai	Sao Paulo	Sao Paulo
London	New York	New York	New York		Mexico City	Mexico City	Bombay
	London	Los Angeles	Los Angeles		Buenos Aires	Shanghai	Shanghai
	Paris	Osaka	Osaka		Beijing	Bombay	Lagos
	Los Angeles	Paris	Moscow		Sao Paulo	Buenos Aires	Mexico City
		Moscow	Paris			Seoul	Beijing
						Rio de Janeiro	Dacca
						Beijing	Jakarta
						Calcutta	Karachi
						Tianjin	Manila
						Jakarta	Tianjin
						Manila	Calcutta
						Cairo	Delhi
						Delhi	Seoul
							Buenos Aires
							Cairo
							Rio de Janeiro
							Bangkok
							Teheran
							Istanbul
							Lima
							Hyderabad
							Lahore
							Madras

Source: UN (1993).

15.2 The challenge

In view of the escalation of dengue infections, particularly dengue haemorrhagic fever, since 1953, it would seem possible that dengue may come to rival malaria as the major vector-borne disease. The World Health Organization (WHO) estimates that worldwide 267 million people are ill with malaria and two million die from it each year (WHO 1992). Four million infants and children die each year from diarrhoeal diseases, largely due to contaminated food and water.

The problem of dengue is particularly acute among the urban poor, i.e. those living in slums and squatter settlements (Figure 15.2). Often they occupy land that is marginal, e.g. favelas on the steep hillsides of Rio de Janeiro, near factories or garbage dumps such as Smokey mountain outside Manila, or around international airports. In the latter case of, course, this merely increases national receptivity to exotic diseases and vectors. For example, the secondary dengue vector *Aedes albopictus* is currently expanding its global range, principally via the importation of containers of used vehicle tyres. Since 1985, *Aedes albopictus* has entered and spread through twenty-four states of the USA, eight states in Brazil, throughout Nigeria and in most of Italy to name a few (Knudsen 1996). Due to the predilection of *Aedes albopictus* for breeding in both artificial and natural habitats such as tree holes, plant axils and bracts, it is well suited to peri-urban environments.

Lee (1994) has analysed urban vector control and town planning in South-East Asian cities. Although rapidly developing dense urban settlement undoubtedly poses a challenge to environmental health, it poses a more basic challenge to most municipal authorities, which must develop the institutional means and basic infrastructure and services. Environmental health problems are worst where the local government structures have not changed to meet demands.

Development of coherent urban land policies will come too late. Very few governments in Asia have developed comprehensive policies that meet intersectoral needs and co-ordinate activities to meet stated objectives (Archer 1990). Rising urban land values represent an obvious and appropriate source of income to finance the necessary infrastructure, but many governments fail because they do not have an effective land valuation and taxation system. Because some governments deem that squatter settlements are illegal, no services are provided, often to the detriment of legal neighbours. It is therefore not surprising that such settlements are characterized by high insecurity of tenure, which results in low investment generally in community infrastructure. This increases the risk of vector-borne diseases such as dengue.

Apart from the development of better planning and tax collection, effective collaboration is required between all levels of government and the community. Anthropologists have now provided frameworks for collecting data on attitudes, beliefs and behaviours of communities in relation to problem recognition and, more importantly, analysis of these data in relation to planning and carrying out interventions (Klein 1992). Put simply, if the people in a community believe that

Figure 15.2 Knocking down slum areas in Singapore and relocating the residents in high rise apartments also provides a form of vector control.

dengue is 'a cramp-like seizure caused by evil spirits', as is implied by the name's putative Swahili derivation, then the public health authorities will have no chance of getting them motivated into anti-mosquito (or anti-litter) activities. At an interdepartmental level, the biologists and non-health professionals such as politicians, administrators, engineers and town planners need to be able to develop a common language and policy to meet mutual objectives. It is important that public health personnel realize that drainage improvements to a neighbourhood are likely to be a more cost-effective and persistent means of controlling mosquito breeding than is constantly using insecticidal sprays.

The community has a clear role to play with respect to the development of desirable infrastructures and services, whether it be co-operation with necessary land acquisition for water supply and drainage installations or in garbage removal.

Non-government organizations and other bodies can provide an important link between the community and government through projects designed to improve community assistance through innovative participatory methods, technical assistance or through income-generating and loans schemes.

In southern Vietnam, Jacques Boisvert and collaborators from the Pasteur Institute, Ho Chi Minh City, provide development loans to villagers who are willing to make covers for their earthenware jars, which are usually infested with *Aedes aegypti* (Boisvert, personal communication). In Kerala, India, Kris Panicker and colleagues have stimulated villagers into forming an 8000-person Filariasis Control Movement (FILCO), which is now a registered charity (Panicker and Dhanda 1992). The problem of Brugian filariasis was especially acute in the Shertallai region of Kerala. *Brugia malayi* causes elephantiasis, which results in gross swelling, especially of the lower limbs and genitalia, leading in turn to social problems. Consequently, in 1986, the Vector Control Research Centre, Pondicherry, India, introduced a community-based strategy there to clear the innumerable ponds, canals and channels of aquatic weeds such as *Pistia*, *Eichhornia* and *Salvinia* (Panicker and Rajagopalen 1990). Larvae of *Mansonia* mosquitoes, which transmit filariasis, are dependent on attaching to the roots of such plants, and clearing the weeds eliminates their breeding habitat. The core of this community-based programme lay with the introduction of fish-rearing as an income-generating exercise. Among the fingerlings distributed were weed-eating edible fish such as the giant gourami (*Osphronemus goramy*) and Chinese grass carp (*Ctenopharyngodon idella*). Because of its success, the National Bank for Agriculture and Rural Development has now taken over the scheme and makes loans to those wishing to make economic use of weed-infested ponds. The scheme has also financed the running of seventy-five filariasis detection and treatment centres staffed by over 200 volunteers.

Similar success was achieved at Pudukuppam, a coastal village in Pondicherry territory, where villagers turned the removal of the filamentous algae *Enteromorpha compressa* into income-generating paper production (Panicker and Rajagopalen 1990). This estuarine alga provided a major breeding site for *Anopheles subpictus*, which is responsible for mesoendemic malaria in the region.

The challenge, therefore, applies to all concerned with the urban and peri-urban environment. The International Drinking Water Supply and Sanitation decade from 1981 to 1990 made notable achievements in providing approximately 1,348 million more people with safe drinking water, and 748 million more people with sanitation (Najlis and Edwards 1991). However, there is much more to be done.

15.3 Strategies against dengue

Much has been made of the relative merits of vertical or 'top-down' and horizontal or 'bottom-up' strategies (Gubler 1989) for the control of the mosquito vectors of dengue (and yellow fever). In the urban context, and even at rural village level, both strategies are mainly directed at *Aedes aegypti*; the former was executed by Soper, Gorgas and Oswald Cruz with full legislative back-up, meticulous administration and house-to-house coverage (Duffy 1977). The yellow flag used even today to mark a site of inspection by public health staff in Brazil is a reminder of the glory days when eradication of *Aedes aegypti* was achieved in nineteen Latin American countries.

Vertical progammes depend on source reduction, removal or modification of potential breeding sites of *Aedes aegypti*, and the knowledge by the public that they will be fined if their households are infested. These programmes were supplemented by the use of insecticides and health education. Whereas some politicians in Queensland, Australia, are loath to accept that the 'pot plant police' approach (as it was called in the press) would do any more than bring repercussions at the ballot box, Singapore has a well accepted programme built on this strategy (Wang 1994). Although payment of fines may be unpalatable to 6,000 people each year, it is totally acceptable to the 600,000 majority who are diligent enough to ensure that they control domestic mosquito breeding.

In 1981, Cuba experienced the first epidemic of dengue haemorrhagic fever in the western hemisphere (Armada and Figueredo 1986). This affected 344,000 people, of whom 116,000 were hospitalized; fortunately the death rate was low, however, at 158. A rigorous vertical campaign was mounted involving fifteen provincial directors, sixty entomologists, twenty-seven general supervisors, 729 team leaders, 3,801 inspectors and 1,847 controllers to clean up or treat household breeding sites. Decree-Law 27, making mosquito breeding illegal, involved nineteen two-monthly inspection cycles in fifteen provinces. Eradication was reported by fourteen.

The high cost of vertical programmes is said to make them unsustainable, and for this reason many people have been promoting the other end of the scale, i.e. the 'bottom-up' or community-based approaches (Kay 1994). Devolvement of responsibility to the community is popular in times of economic rationalism, but in reality this approach is unlikely to succeed unless there are financial rewards or perceived tangible benefits. After all, that was the crux of the Indian examples at Shertallai and Pudukuppam.

With respect to dengue vector control, the Rockefeller Foundation has been heavily committed towards developing community-based programmes in Mexico and Honduras (Halstead and Gomez-Dantes 1992) with limited success. In St Lucia, Dominica and Trinidad, larval densities returned to pre-project levels within months of completion (Nathan 1992), and in Ceara, Brazil, villagers promptly downgraded use of predacious copepods (see below) against *Aedes aegypti* breeding in water storage units when cholera threatened (Vasconcelos *et al.* 1992) (Figure 15.3). In Thailand, distribution of larvicide at village level was diverted for purposes other than dengue vector control (Boonluan *et al.* 1986).

Fortunately there is now at least one example of how community-based *Aedes aegypti* control can work. In 1989, as a WHO consultant, I introduced Vietnamese health personnel to the concept of using fresh water microcrustaceans called copepods, genus *Mesocyclops*, as predators of newly hatched larvae. Vietnam is

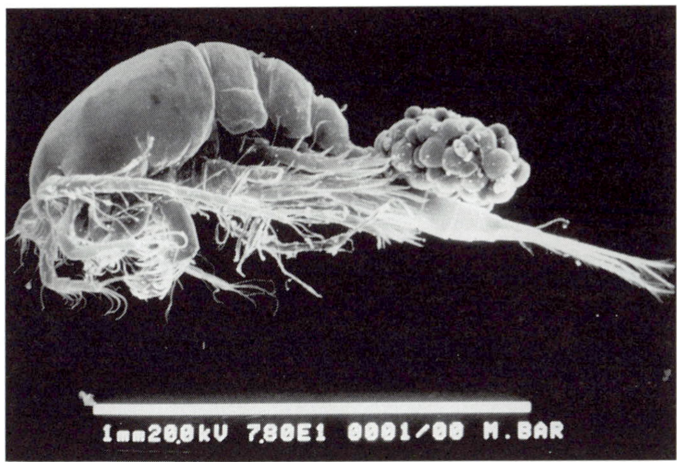

Figure 15.3 Predacious microcrustaceans called copepods are approximately 1 mm long and live in fresh water. The genus *Mesocyclops*, in particular, has been very useful in biological control programmes against dengue mosquitoes.

copepod heaven because of its multitude of eutrophic ponds, lakes and pools that are polluted with green-coloured unicellular algae, among other things. Local *Mesocyclops* species were collected and evaluated in the National Institute of Hygiene and Epidemiology laboratory, run by Dr Vu Sinh Nam (Nam *et al.*, 1998). After consultation with the commune leader, Phanboi and Nhanvinh villages, 31 km east of Hanoi, were selected for a pilot trial to see whether locally collected copepods would establish in outdoor concrete tanks, wells and ceramic jars.

Pilot releases of mixed copepod species including *Mesocyclops woutersi*, *Mesocyclops thermocyclopoides* and *Mesocylcops ruttneri* during February and June 1993 indicated sufficient promise to invite the community to take an active role. The principles of using predacious *Mesocyclops* were discussed with village leaders, health workers and the youth and women's union representatives. This information was conveyed to all householders who, during March 1994, inoculated all containers in Phanboi, again using previously established stocks. Nhanvinh was kept as an untreated control as part of the scientific evaluation process. Since August 1994, no *Aedes aegypti* larvae have been detected, and, except for a single individual collected resting inside a house in November 1994, no adults have been collected either.

An important component to enhance the introduction of *Mesocyclops* was intensification of the recycling carried out by families to supplement their income, which is mainly from rice growing. It is common to see villagers on bicycles peddling towards communal redistribution centres with large hemp sacks of bottles, tins, tyres, battery cases or discarded plastic items draped over the rear luggage rack. Women and children are often seen washing and drying used plastic bags on roadside allotments. When the health implications in terms of reducing the number of potential small breeding sites for *Aedes aegypti* was explained to them, some villagers were bemused. After all, they were doing it for supplementary income; the health issue was simply a bonus. However, without this whole-hearted co-operation, eradication would not have been achieved. This is the first time that contemporary horizontal methodologies have been used successfully to achieve dengue eradication. The prospects for many other parts of Vietnam and other countries where the problem of mosquito breeding in large water storage containers (which are not washed regularly) seem exceedingly bright.

Predacious copepods are not a panacea, and successful programmes in dengue control will have to be tailored according to the key sites of *Aedes aegypti* production and social and cultural norms. Whether they are vertical, horizontal or oblique (government supervised community action) does not really matter as long as they effectively control the breeding of dengue mosquitoes.

References

Archer, R.W. (1990) *An outline urban land policy for the developing countries of Asia.* Research Paper No. 20 for Human Settlements Division, Asian Institute of Technology, Bangkok.

Armada, J.A.G. and Figueredo, R. (1986) Application of environmental management principles in the program for eradication of *Aedes (Stegomyia) aegypti* (Linnaeus 1762) in the republic of Cuba. *Bulletin of the Pan American Health Organization*, **20**, 186–193.

Batra, C.P. and Reuben, R. (1979) Breeding of *Anopheles stephensi* (Liston) in wells and cisterns in Salem, Tamil Nadu. *Indian Journal of Medical Research*, **70** (supplement), 114–122.

Boonluan, P., Wirat, S., Prakong, P.U., Somkiat, B. and Banyong, M. (1986) *Approaches for community participation in* Aedes aegypti *control, Phanus Nikhom district, Chonburi province.* Proceedings of the ICMR/WHO Workshop to Review Research, Malaria Research Centre, Delhi.

Cairncross, S. (1990) Water Supply and the Urban Poor. In: *The Poor Die Young: Housing and Health in Third World Cities* (eds Hardoy, J.E., Cairncross, S. and Satterthwaite, D.). Earthscan Publications, London, pp. 109–126.

Cairncross, S. and Feachem, R.G. (1983) *Environmental health engineering in the tropics – an introductory text.* John Wiley and Sons, Chichester.

Duffy, J. (ed.) (1977) *Ventures in World Health.* Pan American Health Organization Scientific Publication No. 355, Washington, DC.

Eshghy, N. (1977) *Anopheles multicolor* and its role in the transmission of malaria in Iran. *Journal of the Entomological Society of Iran*, **4**, 17–22.

Gubler, D.J. (1989) *Aedes aegypti* and *Aedes aegypti*-borne disease control in the 1990s: top down or bottom up. *American Journal of Tropical Medicine and Hygiene*, **40**, 571–578.

Halstead, S.B. and Gomez-Dantes, H. (eds) (1992) *Dengue: a Worldwide Problem, a Common Strategy.* Ministry of Health, Mexico, and Rockefeller Foundation, New York.

Jennings, C.D. Phommasack, B., Sourignadeth, S. and Kay, B.H. (1995) *Aedes aegypti* control in the Lao People's Republic, with reference to copepods. *American Journal of Tropical Medicine and Hygiene*, **53**, 324–330.

Kay, B.H. (1994) Intersectoral approaches to dengue vector control. *Kaohsiung Journal of Medical Sciences*, **10**, 556–561.

Kay, B.H., Cabral, C.P., Araujo, D.B., Ribeiro, Z.M., Braga, P.H. and Sleigh, A.C. (1992) Evaluation of a funnel trap for the collection of copepods and immature mosquitoes from wells. *Journal of the American Mosquito Control Association*, **8**, 372–375.

Klein, R.E. (1992) Quantitative and qualitative methods in the support of health interventions. In: *Dengue: a Worldwide Problem, a Common Strategy* (eds Halstead, S.B. and Gomez-Dantes, H.). Ministry of Health, Mexico, and Rockefeller Foundation, New York, pp. 95–105.

Knudsen, A.B. (1992) Towards a global strategy for the control of dengue and dengue haemorrhagic fever. In: *Dengue: a Worldwide Problem, a Common Strategy* (eds Halstead, S.B. and Gomez-Dantes, H.). Ministry of Health, Mexico, and Rockefeller Foundation, New York, pp. 49–53.

Knudsen, A.B. (1996) Distribution of vectors of dengue fever/dengue haemorrhagic fever with special reference to *Aedes albopictus*, 1996. *Dengue Bulletin*, **20**, 5–12.

Lee, Y.F. (1994) Urban planning and vector control in southeast Asian cities. *Kaohsiung Journal of Medical Sciences*, **10**, 539–551.

Najlis, P. and Edwards, A. (1991) The international drinking water supply and sanitation decade in retrospect and implications for the future. *Natural Resources Forum*, **15**, 110–117.

Nam, V.S., Yen, N.T., Kay, B.H., Marten, G.G. and Reid, J.W. (1998) Eradication of *Aedes aegypti* from a village in Vietnam using copepods and community participation. *American Journal of Tropical Medicine and Hygiene* (in press).

Nathan, M.B. (1992) An overview of dengue prevention and control programs in the English speaking Caribbean. In: *Dengue: a Worldwide Problem, a Common Strategy* (eds Halstead, S.B. and Gomez-Dantes, H.). Ministry of Health, Mexico, and Rockefeller Foundation, New York, pp. 207–210.

Organization for Economic Co-operation and Development (1991) *The State of the Environment*. OECD, Paris.

Panicker, K.N. and Dhanda, V. (1992) Community participation in the control of filariasis. *World Health Forum*, **13**, 177–179.

Panicker, K.N. and Rajagopalen, P.K. (1990) *A success story of community participation in malaria control*. Miscellaneous publications of the Vector Control Research Centre, Pondicherry, No. 18.

Panicker, K.N., Geetha Bai, M. and Kalyanasundaram, M. (1982) Well breeding behaviour of *Aedes aegypti*. *Indian Journal of Medical Research*, **76**, 689–691.

Russell, B.M., Muir, L.E., Weinstein, P. and Kay, B.H. (1996) *Aedes aegypti*, wells and gold mines in northern Australia – surveillance and control. *Medical and Veterinary Entomology*, **10**, 155–160.

Sinnatamby, G. (1990) Low cost sanitation. In: *The Poor Die Young: Housing and Health in Third World Cities* (eds Hardoy, J.E., Cairncross, S. and Satterthwaite, D.), Earthscan Publications, London, pp. 127–167.

United Nations (1993) *World Urbanization Prospects*, the 1992 revisions. UN, New York.

Vasconcelos, A.W., Sleigh, A.C., Kay, B.H., Cabral, C.P., Araujo, D.B., Ribeiro, Z.M.,

Braga, P.H. and Cavalcante, J.S., Jr (1992) Community use of copepods to control *Aedes aegypti* in Brazil. In: *Dengue: a Worldwide Problem, a Common Strategy* (eds Halstead, S.B. and Gomez-Dantes, H.). Ministry of Health, Mexico, and Rockefeller Foundation, New York, pp. 139–144.

Wang, N.C. (1994) Control of dengue vectors in Singapore. *Kaohsiung Journal of Medical Science*, **10**, 533–538.

World Health Assembly (1993) *Dengue Prevention and Control*. WHA 46.31, Geneva.

World Health Organization (1992) *Our Planet, our Health*. Report of the WHO Commission on Health and Environment, Geneva.

Yasuno, M., Rajagopalen, P.K., Kazmi, S.J. and La Brecque, G.C. (1977) Seasonal changes in larval habitats and population density of *Culex fatigans* in Delhi villages. *Indian Journal of Medical Research*, **65**, 52–64.

Index

Page numbers in *italics* indicate figures or tables